COMPUTATIONAL GEOMETRY FOR SHIPS

Malcolm Bloor is Professor of Mathematical Engineering at the University of Leeds and is currently Head of the Department of Applied Mathematical Studies. He obtained his B.Sc. and Ph.D. from the University of Manchester and has research interests in computer-aided geometric and functional design, and various areas of fluid mechanics.

Christopher Wojciech Dekanski was born in Great Britain in 1968. He graduated from the University of Leeds with a B.Sc. Honours degree in Mathematics in 1989 and obtained a Ph.D. from the Department of Applied Mathematical Studies, University of Leeds, in 1993. He is currently a postdoctoral research fellow in the Department of Applied Mathematical Studies. His research interests include computer-aided geometric and functional design, and computational fluid dynamics.

Jan Piotr Michalski was born in Poland in 1942. He was educated at the Technical University of Gdansk (Shipbuilding Faculty), where he obtained an M.Sc. in 1966 and a Ph.D. in 1974. His field of specialization is computer-aided ship design. He is now at the Technical University of Gdansk.

Horst Nowacki was born in Berlin in 1933. He was educated as a naval architect, receiving his doctorate from the Technical University of Berlin in 1963. He held the post of Professor of Naval Architecture at the University of Michigan from 1964–74, and has been Professor of Ship Design at the Technical University of Berlin since 1974.

Boguslaw Oleksiewicz was born in Poland in 1944. He was educated at the Technical University of Gdansk (Shipbuilding Faculty), where he obtained an M.Sc. in 1968 and a Ph.D. in 1985. He specializes in computer-aided ship design and is currently at the Technical University of Gdansk.

Michael John Wilson graduated from the University of Cambridge with a B.A. Honours degree in Natural Science in 1980. He obtained a Ph.D. from the Cavendish Laboratory in Cambridge in 1984. He moved to the Department of Applied Mathematical Studies at Leeds University as a postdoctoral research fellow and was eventually appointed as a lecturer in 1986. He is currently a Senior Lecturer in Applied Mathematics. His research interests include computer-aided geometric and functional design, computational fluid dynamics, industrial mathematics, and astrophysics.

COMPUTATIONAL GEOMETRY FOR SHIPS

Editors

H. Nowacki
Technical University of Berlin

M.I.G. Bloor
University of Leeds

B. Oleksiewicz
Technical University of Gdansk

World Scientific
Singapore • New Jersey • London • Hong Kong

Published by

World Scientific Publishing Co. Pte. Ltd.

5 Toh Tuck Link, Singapore 596224

USA office: 27 Warren Street, Suite 401-402, Hackensack, NJ 07601

UK office: 57 Shelton Street, Covent Garden, London WC2H 9HE

Library of Congress Cataloging-in-Publication Data
Computational geometry for ships / editors, H. Nowacki, B.
   Oleksiewicz, M. I. G. Bloor.
     p. cm.
   Includes bibliographical references (p.   ) and index.
   ISBN-13 978-981-02-2139-3 (hard cover) -- ISBN-10 981-02-2139-8 (hard cover)
   ISBN-13 978-981-02-3353-2 (pbk) -- ISBN-10 981-02-3353-1 (pbk)
   1. Naval architecture--Data processing. 2. Hulls (Naval architecture)--Design and construction--
Data processing. 3. Shipbuilding--Data processing. 4. Geometry--Data processing.
I. Nowacki, H. (Horst) II. Oleksiewicz, B. III. Bloor, M. I. G.
   VM156.C546 1995
   623.8'4--dc20
                                                              94-46516

British Library Cataloguing-in-Publication Data
A catalogue record for this book is available from the British Library.

Cover illustration by Dipl. Ing. Stefan Harries MSE.

Contributing Authors

H. Nowacki
Technical University of Berlin

J. Michalski   B. Oleksiewicz
Technical Univesity of Gdansk

M. I. G. Bloor   C. W. Dekanski   M. J. Wilson
University of Leeds

Sponsored by:

The European Commission
TEMPUS Project JEP 3051

vii

PREFACE

This book is based on the results of a Joint European Project (JEP 3051) sponsored by the Tempus Office of the European Commission. This project brought together six authors from three European universities, the Technical University of Gdansk, the Technical University of Berlin, and the University of Leeds who jointly developed the material for an advanced course on the Computational Geometry for Ships. The course was initiated at the Technical University of Gdansk where it was given three times between 1993 and 1995. The book contains the essential substance of this course and thus documents a systematic and comprehensive approach to the definition of ship hull geometry.

The book is thus intended to serve as an introduction to and as a reference for the mathematical fundamentals of hull form definition and for modern computational methods of hull shape design. This includes a description of the methods for ship geometry generation in initial design, shape quality improvement by fairing processes, and surface quality interrogation and evaluation. The curves and surfaces of the hull geometry are represented by Bézier, B-Spline, Hermite, NURBS and other geometric bases. These methods form the foundations for most modern computing systems used in defining the free-form shapes of ships and many other complex technical shapes. Many of these methods are, for instance, also applied to the description of car body and aircraft shapes and in similar sister disciplines.

The authors acknowledge their appreciation and gratitude to the TEMPUS program for the opportunity to document the state of the art in the fast advancing field of computational geometry for ships.

For the editors and authors:

Horst Nowacki

viii

ACKNOWLEDGEMENTS

The authors would like to thank the administrative and technical support given by Detlef Schulze and Wojtek Puch and for the hospitality provided by the Technical University of Gdansk during the running of the course.

CONTENTS

CHAPTER 1

INTRODUCTION

1. History

The design and representation of ship hull forms by means of mathematical descriptions has a long tradition. Simple geometric shapes such as circles, ellipses, parabolas or sine curves can be found in ancient and medieval ships. Such simple shapes are easy to manufacture and reproduce.

Ship lines plans as a means to develop and document empirical, free-form hull shapes were introduced around 1700 together with lofting and draughting methods using elastic splines as a fairing tool. The results of this technology were remarkable; it produced successful ship geometries that could be reliably manufactured. The mathematical representation of these empirically developed lines was not of interest in the beginning.

The Swedish naval constructor Chapman in his famous book "A Treatise on Ship-Building" [14] around 1760 mentions the use of a family of parabolas for waterlines and other ship curves. Later pioneering work on mathematical ship lines was performed by D. W. Taylor [74] right after the turn of this century who began to use mathematical functions to describe the empirical hull shapes of his era. He presented the sectional area curve and waterlines of the ship by explicit polynomials of degree 5

$$y(x) = \sum_{i=0}^{5} a_i x^i. \tag{1}$$

This enabled him to specify and systematically vary six free form parameters for a half-body curve segment of this type, e.g:

- Offsets at segment endings

- Slopes at segment endings

- Curvature, say, amidships

- Area under the curve (prismatic coeff.)

From this approach Taylor developed the parent hull shapes which were investigated in his classical systematic hull form series (Taylor series). He was thus the first naval architect to use mathematical representations for the systematic variation of hull shape by a family of form parameters. This approach was later adopted by many others in systematic investigations.

Saunders [68] gives a good account of the early history of mathematical ship lines. Weinblum [80] summarizes the status achieved in systematic hull form development in the fifties.

The advent of the computer in shipbuilding, also since the early fifties, changed the scope, style and methodology of mathematical ship form definition in a dramatic way. Today modern shipbuilding around the world can no longer be imagined without intensive use of computers and computational methods of ship geometry definition. New techniques for ship geometry design are supported by computer today. The production of ship parts is performed by numerically controlled machine tools which require accurate, computer based ship geometry information as input. Numerous software systems exist to develop ship geometry design and production information.

2. Objectives

In recent decades, mathematical methods of computational geometry have made much progress and are successfully applied in many engineering disciplines. This course will give an introduction into modern computational methods for ship geometry. It will review the mathematical background needed to understand the methods used by modern computer systems for ship geometry design and manufacture.

The material in these notes will focus on several lines of development which stem from different objectives and motivations and will lead to a variety of applications. The main perspectives include:

- Lines plan definition:
  procedures based on curve generation methodology close to manual naval architecture practice.
  Application of spline fairing methods.

- Ship surface definition:
  methods based on curve mesh fairing and approximation. Shape representation by Coons, Bézier, and B-Spline methods.

- Form parameter design:
  direct shape generation from specified form parameters.

- Shape optimization:
  methods for improving shape elements on the basis of explicit shape quality criteria.

- Shape evaluation:
  methods providing a feedback for geometric design decisions from computational evaluation of functional properties of shape.

3. CAD Systems

Computer aided design systems are intended to serve the human designer in product design tasks. They require a well-conceived division of labour between man and machine. They do not aspire to automate design work, so they must rely on man-machine interaction. A natural division of roles between man and computer corresponds to the three development stages of a product:

1 Conceptual stage, idea stage:
   Creative work, human task
2 Evaluation stage:
   Calculation and analytical work, computer task
3 Judgement stage:
   Critical work, human task

CAD systems have two main objectives:

- To support the product design process

- To produce a computer internal product definition

The design process is a structured decision process from a set of requirements to a product solution meeting technical and economic goals. It is characterized by the following elements (Fig. 1):

D = decision variables (free variables, under the designer's control)

P = parameters (variable inputs, not under the designer's control)

M = measure of merit (a criterion for the principal design objective)

C = constraints (side conditions, equalities and inequalities, the product must meet)

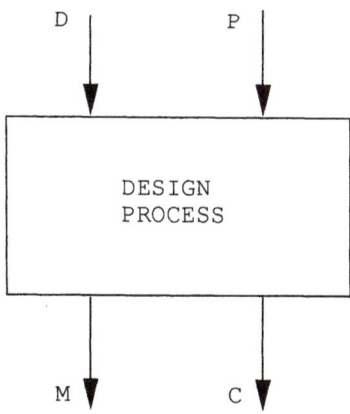

Fig. 1. Model of design process

In the geometric design of a ship shape element, e.g., the free form parameters of the shape are design variables, a fairness measure may be chosen as measure of merit or end conditions may act as constraints.

It is the designer's task to formulate the design problem as a structured decision process. The CAD system's role is to support this process in its steps and to document the resulting product in its database. A suitable, unique data structure is needed to store the product definition.

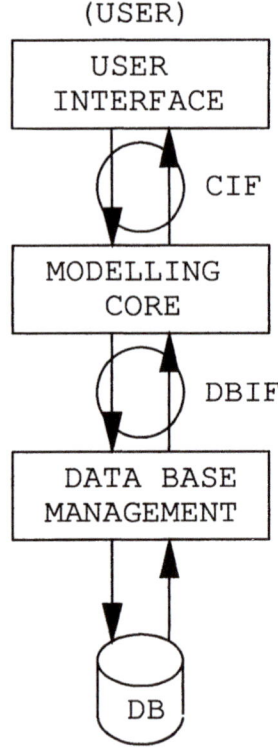

Fig. 2. Basic software architecture of CAD system

A CAD system has the following basic structure (architecture) of its software components (Fig. 2):

- User interface module:
    Supporting the interaction between user and system by alphanumeric input / output, command and/or menu interpretation, execution control for CAD tasks.

- Modelling core:
    Core of product modelling functions creating elements of the computer internal

product model and interrogating the model for calculation and analysis purposes.

- Database management:
  Management of DB access (storage / retrieval) based on the chosen product data structure.

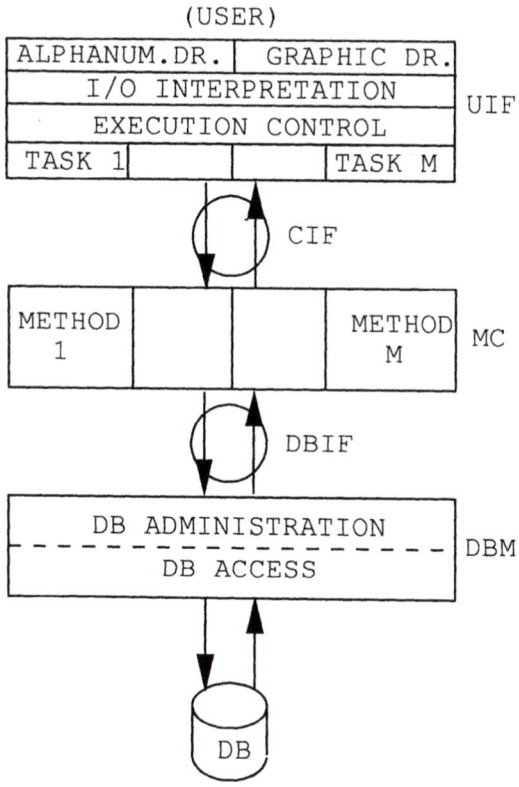

Fig. 3. Refined view of system architecture

Fig. 3 shows a refined view of this basic system architecture. The software functions are subdivided by modules and layers as follows:

- Alphanumeric and graphic device driver routines (device dependent)

- Input / output (I/O) interpretation, e.g., command interpretation or menu routines for tablet / screen

- Execution control of CAD system, task supervisor, error handling etc.

- Task control,
  supervisor for tasks which refer to methods, task error handling

- CIF:
  Core interface, procedural interface to core functions

- Modelling core (MC):
  Executable methods under their own control, modelling and interrogation functions, analysis functions, free of I/O functionality and DB functions

- DBIF:
  Database interface, procedural interface to DB management

- Database Management (DBM):
  Administrative functions for database management
  DB access functions

These functions can be realized in many different ways. In the context of systems for ship geometry definition the modelling process focuses on functions for geometry definition, manipulation, and interrogation. The contents of this book concentrate on those mathematical and computational methods which form the modelling core for these systems. In addition users of computational geometry systems for ships need to become familiar with the user interface and database administration in their system environment. The background gained from this book on the possible approaches to ship geometry definition will provide valuable guidance to practitioners who wish to understand the mathematical basis of modern computer systems for ship geometry.

CHAPTER 2
CURVE DEFINITION

1. Introduction

The purpose of this chapter is to introduce some elementary techniques for the mathematical description of curves. The ideas described here will be used in those chapters that follow which are concerned with the description of hulls by curves, and will be built upon in those chapters concerned with the surface definition of hulls. In this chapter we will consider only Cartesian coordinate systems.

2. Parametric Curve Representation

A familiar way of representing a plane curve is by using explicit non-parametric equations of the form $y = f(x)$. However, in CAGD it is more common to define curves parametrically in terms of a single scalar parameter. There are a number of reasons for this: for one thing it makes the curves easier to manipulate, while for the purposes of this chapter it makes the mathematical properties of the curves easier to evaluate.

Consider a curve $\mathbf{r}(t)$ defined by a vector-valued function of the parameter t by

$$\mathbf{r}(t) = (r_1(t), r_2(t), r_3(t)) \tag{2}$$

where the scalar functions $r_1(t), r_2(t), r_3(t)$ are called the coordinate functions of $\mathbf{r}$ and give the x, y and z coordinates of points along the line.

Thus we define a parametrized differentiable curve as a differentiable function $\mathbf{r}$ and can regard it as a mapping from the real-line R to Euclidean 3-space E^3 (Fig. 4). In CAGD, it is usual to consider a limited region of the real line, e.g. $0 \leq t \leq 1$.

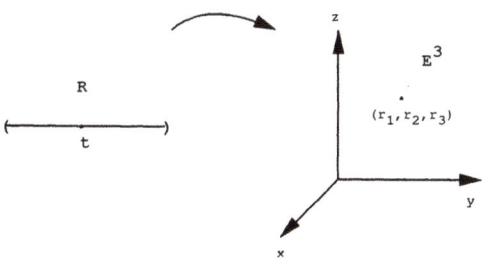

Fig. 4. The mapping from R into E^3

We further say that a curve $\mathbf{r}(t)$ of a real variable t is smooth (or differentiable) if its derivatives $\frac{dr_1}{dt}, \frac{dr_2}{dt}$ and $\frac{dr_3}{dt}$ all exist [13].

Examples

o The parametrized differentiable curve given by $r(t) = (a \cos t, a \sin t, bt)$ where t is real represents a helix of fixed pitch $2\pi b$ on the cylinder $x^2 + y^2 = a^2$ in E^3. The parameter t measures the angle which the x-axis makes with the line joining the origin O to the projection of the point $r(t)$ over the (x, y) plane.

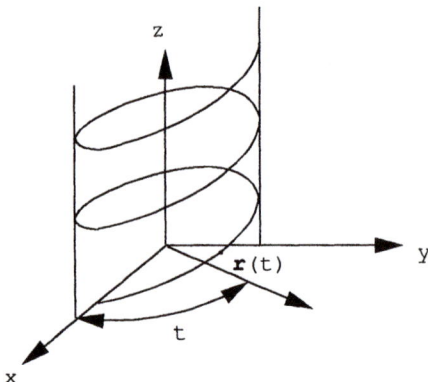

Fig. 5. The helix $r(t) = (a \cos t, a \sin t, bt)$

o The function $r(t) = (t, |t|)$ is not a differentiable curve, since $|t|$ is not differentiable at $t = 0$.

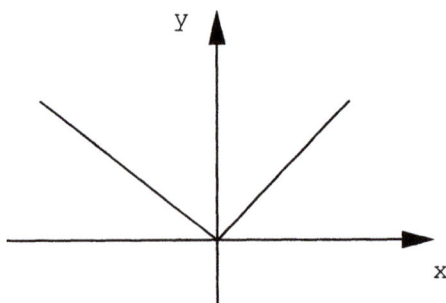

Fig. 6. $r(t) = (t, |t|)$

2.1. The Velocity Vector for a Curve

If $\mathbf{r}(t) = (r_1(t), r_2(t), r_3(t))$ is a curve in E^3 then the velocity vector of $\mathbf{r}(t)$ is the vector

$$\mathbf{r}' = \left(\frac{dr_1}{dt}, \frac{dr_2}{dt}, \frac{dr_3}{dt}\right). \tag{3}$$

Mathematically

$$\mathbf{r}' = \frac{d\mathbf{r}}{dt} = \lim_{\Delta t \to 0}\left(\frac{\mathbf{r}(t + \Delta t) - \mathbf{r}(t)}{\Delta t}\right). \tag{4}$$

So as $\Delta t \to 0$ the vector $\mathbf{r}(t + \Delta t) - \mathbf{r}(t)$ becomes tangent to the curve at the point $\mathbf{r}(t)$ (Fig. 7).

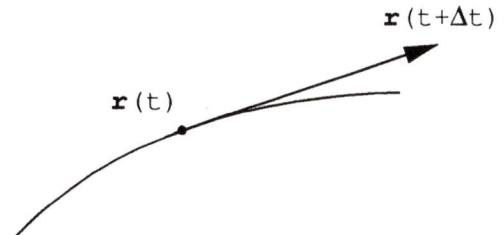

Fig. 7. The tangent vector

2.2. Definition of a Regular Curve

If $\mathbf{r}(t)$ is a parametrized differentiable curve, then for each t where $\mathbf{r}' \neq 0$ there is a well defined straight line tangent to the point $\mathbf{r}(t)$. At points where $\mathbf{r}' = 0$ then a singular point exists. From this we define a parametrized differentiable curve to be regular if $\mathbf{r}' \neq 0$ for all real values of t.

Examples

○ The function $\mathbf{r}(t) = (t^3, t^2)$ is a differentiable curve. However at $t = 0$ there is a cusp and $\mathbf{r}'(0) = (0, 0)$ and so we say that the curve is not regular (Fig. 8). Therefore we have a geometric definition of a parametrized differentiable regular curve $\mathbf{r}(t)$. We will be dealing in the main with regular curves throughout this work. In CAGD it is very common for curves to be specified as polynomial functions of the parameter t, where it is convenient for t to lie in the range $0 \leq t \leq 1$.

○ Consider

$$\mathbf{r}(t) = \left((1 - t^3)\mathbf{a_0} + 3t(1 - t)^2\mathbf{a_1} + 3t^2(1 - t)\mathbf{a_2} + t^3\mathbf{a_3}\right) \tag{5}$$

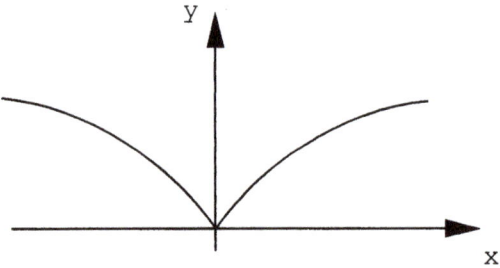

Fig. 8. The cusped curve $\mathbf{r}(t) = (t^3, t^2)$

where $\mathbf{a_0}, \mathbf{a_1}, \mathbf{a_2}, \mathbf{a_3}$ are vector constants or control points of the curve. This is the definition of a Bézier cubic curve [47] which is widely used in CAGD. The velocity vector for this curve is given by

$$\mathbf{r'} = 3(1 - t^2)(\mathbf{a_1} - \mathbf{a_0}) + 6t(1 - t)(\mathbf{a_2} - \mathbf{a_1}) + 3t^2(\mathbf{a_3} - \mathbf{a_2}) . \qquad (6)$$

Note that for

$$t = 0 \quad \mathbf{r'} = 3(\mathbf{a_1} - \mathbf{a_0}) \text{ which is parallel to } (\mathbf{a_1} - \mathbf{a_0})$$
$$t = 1 \quad \mathbf{r'} = 3(\mathbf{a_3} - \mathbf{a_2}) \text{ which is parallel to } (\mathbf{a_3} - \mathbf{a_2})$$

Thus we can see that by moving the control points $\mathbf{a_1}$ and $\mathbf{a_2}$ we can alter the tangent directions at the end positions, and hence alter the shape of the curve (Fig. 9).

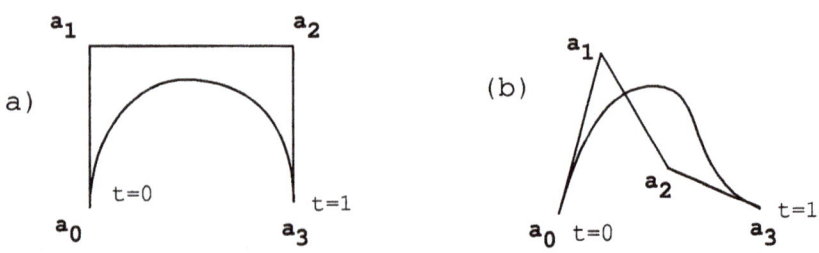

Fig. 9. A Bézier curve and control points

2.3. Reparametrization of Curves

If we consider the two curves

$$\mathbf{r}(t) = (\cos t, \sin t, 0) \tag{7}$$
$$\mathbf{q}(s) = (\cos 2s, \sin 2s, 0) \tag{8}$$

we can see that these both give rise to the same path in space (namely the circle $x^2 + y^2 = 1$). They have velocity vectors given by

$$\mathbf{r}' = (-\sin t, \cos t, 0) \tag{9}$$
$$\mathbf{q}'(s) = (-2\sin 2s, 2\cos 2s, 0) \tag{10}$$

and so at a given point in space the velocity vector of $\mathbf{q}(s)$ has twice the magnitude of $\mathbf{r}(t)$. However the directions are the same, which is hardly surprising since they are both tangent to the same path in space.

Now if I and J are two intervals on the real line R (Fig. 10) and $\mathbf{r}(t)$ describes a curve, then if we define a differentiable function $h(s) = t$ we say that the composite function

$$\mathbf{q}(s) = \mathbf{r}(h(s)) \tag{11}$$

is a curve called the reparametrization of $\mathbf{r}(t)$. Furthermore the velocity vector of $\mathbf{q}(s)$ is related to the velocity vector of $\mathbf{r}(t)$ by the formula

$$\mathbf{q}'(s) = \mathbf{r}'\frac{dh}{ds} \tag{12}$$

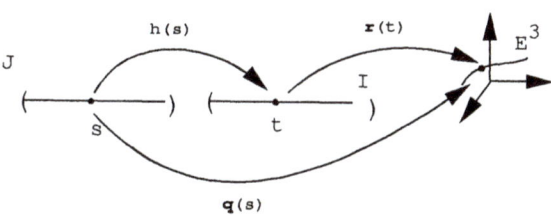

Fig. 10. Curve reparametrization

Example

Suppose $\mathbf{r}(t) = (\sqrt{t}, t\sqrt{t}, 1 - t)$ on the interval $I:(0 < t < 4)$. If $h(s) = s^2$ on $J:(0 < s \leq 2)$ then

$$\mathbf{q}(s) = \mathbf{r}(h(s)) = \mathbf{r}(s^2) = (\sqrt{s^2}, s^2\sqrt{s^2}, 1 - s^2) = (s, s^3, 1 - s^2)$$

2.4. Arc Length of Curve

The length of arc along a curve $\mathbf{r}(t)$ from a reference point $\mathbf{r}(t_0)$ is given by the arc length function

$$s(t) = \int_{t_0}^{t} |\mathbf{r}'(u)| \, du \tag{13}$$

where

$$|\mathbf{r}'(u)| = \sqrt{(x'(u))^2 + (y'(u))^2 + (z'(u))^2} \tag{14}$$

is the speed of the curve.

Since, by assumption $\mathbf{r}'(t) \neq 0$ the arc length s is a differentiable function of t and so

$$\frac{ds}{dt} = |\mathbf{r}'| . \tag{15}$$

3. Arc Length Parametrization

One important way to represent a curve parametrically, which has obvious geometric significance, is to parametrize it in terms of the arc length along it [26]. In general it is not necessary to define the reference point since most important quantities are defined in terms of the derivatives of $\mathbf{r}(s)$.

If $s(t)$, the arc length function, gives a measure of the distance along a curve for a value of t, then the inverse function $t(s)$ is a function which, given the distance along the curve to a point, gives the appropriate value of t at that point. Therefore we can form a reparametrization of $\mathbf{r}$ using $s(t)$; thus $\mathbf{q}(s) = \mathbf{r}(t(s))$, where $t(s)$ is the inverse function of $s(t)$.

Now from Eq. 12

$$\mathbf{q}' = \frac{d\mathbf{q}}{ds} = \frac{dt}{ds}\mathbf{r}' \tag{16}$$

and so, by Eq. 15

$$\mathbf{q}' = \frac{\mathbf{r}'}{|\mathbf{r}'|}. \tag{17}$$

Thus $\mathbf{q}'$ is a unit vector,i.e.

$$|\mathbf{q}'| = \frac{|\mathbf{r}'|}{|\mathbf{r}'|} = 1. \tag{18}$$

In other words a curve parametrized by arc length has unit speed.

Example

Consider the curve

$$\begin{aligned}
\mathbf{r}(t) &= (a \cos t, a \sin t, bt) \\
\mathbf{r}' &= (-a \sin t, a \cos t, b)
\end{aligned}$$

$$\Rightarrow |\mathbf{r}'|^2 = (-a \sin t)^2 + (a \cos t)^2 + b^2 = a^2 + b^2 = c^2$$

$$\Rightarrow |\mathbf{r}'| = c$$

Therefore the arc length function for this curve is given by

$$s(t) = \int_0^t |\mathbf{r}'(u)| du = \int_0^t c \, du = ct$$

$$\Rightarrow s(t) = ct \text{ and } t(s) = s/c$$

$$\Rightarrow \mathbf{q}(s) = \mathbf{r}(t(s)) = (a \cos(s/c), a \sin(s/c), bs/c)$$

and

$$|\mathbf{q}'(s)| = \frac{a^2 + b^2}{c^2} = 1$$

3.1. Orientation

If we are given a curve $\mathbf{r}$ parametrized by arc length s over the range (a, b) we may consider the curve $\mathbf{q}$ defined in $(-b, -a)$ by $\mathbf{q}(-s) = \mathbf{r}(s)$. This has the same path as $\mathbf{r}(s)$ but is described in the opposite sense. Therefore we say that these two curves differ by a change of orientation.

3.2. The Tangent Vector to the Curve

If $\mathbf{q}(s)$ is a curve parametrized by arc length such that $|\mathbf{q}'(s)| = 1$ at every point on the curve then we define the tangent vector by

$$\mathbf{t} = \mathbf{q}'(s) \tag{19}$$

3.3. Curvature

Consider a curve $\mathbf{q}(s)$ parametrized by arc length. Since the tangent vector $\mathbf{q}'(s) = \mathbf{t}$ has unit length, the vector $\mathbf{t}'$ is perpendicular to the tangent of the curve and so must be a normal vector to the curve (Fig. 11). This is because

$$\mathbf{t} \cdot \mathbf{t} = 1 \tag{20}$$

which when differentiated with respect to arc length s gives

$$\mathbf{t}' \cdot \mathbf{t} + \mathbf{t} \cdot \mathbf{t}' = 0 \tag{21}$$

$$\Rightarrow \mathbf{t}' \cdot \mathbf{t} = 0. \tag{22}$$

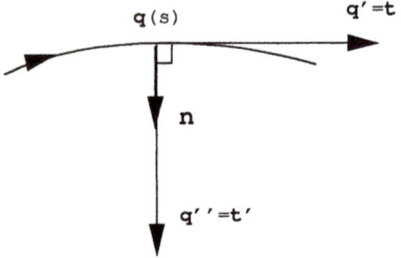

Fig. 11. The curvature vector

t' gives a measure of how rapidly the curve pulls away from the tangent lines at s and is given by

$$t' = k\mathbf{n} \qquad (23)$$

where the value $k(s) = |t'|$ is called the curvature of $\mathbf{q}$ at s and $\mathbf{n}$ is termed the principal normal vector and is given by

$$\mathbf{n} = \frac{t'}{|t'|}. \qquad (24)$$

The reciprocal of the curvature is denoted by

$$\rho = \frac{1}{|k|} \qquad (25)$$

and is called the radius of curvature at $\mathbf{q}(s)$. A point on the curve where the curvature vector $k = 0$ is called a point of inflection [73] and here the radius of curvature is infinite.

3.4. The Binormal Vector and Torsion

The binormal vector $\mathbf{b}(s)$ is defined by

$$\mathbf{b} = \mathbf{t} \times \mathbf{n} \qquad (26)$$

so that $|\mathbf{b}| = 1$, as $|\mathbf{t}| = |\mathbf{n}| = 1$. This implies that at any point on a curve $\mathbf{t}, \mathbf{n}, \mathbf{b}$ form an orthogonal set of vectors in E^3.

Secondly we can show that $\mathbf{b}'$ is parallel to the principal normal vector

$$\mathbf{b}' = -\tau\mathbf{n} \qquad (27)$$

where $\tau = \tau(s)$ is a scalar function of s called the torsion of the curve. The torsion measures the rate of twisting of a curve in space.

4. The Serret Frenet Formulae

The Serret Frenet formulae [73] hold a central position in the theory of curve definition. The formulae were originally derived by the French mathematician Frenet in 1847 and were first published in the form shown below by Serret in 1851. They are as follows:

If $\mathbf{q}(s)$ is a regular smooth unit speed curve parametrized by arc length, with $k(s) > 0$ and torsion $\tau(s)$ then the Serret Frenet formulae express the derivatives of $\mathbf{t}, \mathbf{n}$ and $\mathbf{b}$ in terms of $\mathbf{t}, \mathbf{n}$ and $\mathbf{b}$ themselves.

$$\mathbf{t}' = k\mathbf{n} \tag{28}$$
$$\mathbf{n}' = -k\mathbf{t} + \tau\mathbf{b} \tag{29}$$
$$\mathbf{b}' = -\tau\mathbf{n} \tag{30}$$

Proof

$\mathbf{t}, \mathbf{n}, \mathbf{b}$ are defined as above and are mutually perpendicular unit vectors.

1. $\mathbf{t}' = k\mathbf{n}$ by definition.

2.

$$\mathbf{b} \cdot \mathbf{t} = 0$$
$$\Rightarrow \quad \mathbf{b}' \cdot \mathbf{t} + \mathbf{b} \cdot \mathbf{t}' = 0$$
$$\Rightarrow \quad \mathbf{b}' \cdot \mathbf{t} + \mathbf{b} \cdot k\mathbf{n} = 0$$

Now
$$\mathbf{b} \cdot \mathbf{n} = 0 \quad \Rightarrow \quad \mathbf{b}' \cdot \mathbf{t} = 0.$$

Also
$$\mathbf{b} \cdot \mathbf{b} = 1 \quad \Rightarrow \quad \mathbf{b}' \cdot \mathbf{b} = 0$$

so $\mathbf{b}'$ is perpendicular to both $\mathbf{t}$ and $\mathbf{b}$, and therefore $\mathbf{b}'$ is parallel to $\mathbf{n}$, and therefore $\mathbf{b}' = -\tau\mathbf{n}$ for some value of τ, the torsion.

3. If we assume
$$\mathbf{n}' = \alpha\mathbf{n} + \beta\mathbf{b} + \gamma\mathbf{t} \tag{31}$$

where α, β and γ are scalars, then taking the scalar product of Eq. 31 with $\mathbf{n}$ gives
$$\mathbf{n}' \cdot \mathbf{n} = \alpha$$

but
$$\mathbf{n} \cdot \mathbf{n} = 1 \quad \Rightarrow \quad \mathbf{n}' \cdot \mathbf{n} = 0 \quad \Rightarrow \quad \alpha = 0.$$

Now taking the scalar product of Eq. 31 with $\mathbf{b}$ we obtain
$$\mathbf{n}' \cdot \mathbf{b} = \beta$$

and

$$\mathbf{n} \cdot \mathbf{b} = 0 \; \Rightarrow \; \mathbf{n}' \cdot \mathbf{b} + \mathbf{n} \cdot \mathbf{b}' = 0$$
$$\Rightarrow \; \mathbf{n}' \cdot \mathbf{b} = -\mathbf{n} \cdot \mathbf{b}' = \mathbf{n} \cdot \tau\mathbf{n} = \tau$$
$$\Rightarrow \; \beta = \tau.$$

And taking the scalar product of Eq. 31 with t gives

$$\mathbf{n}' \cdot \mathbf{t} = \gamma$$

and

$$\mathbf{n} \cdot \mathbf{t} = 0 \; \Rightarrow \; \mathbf{n}' \cdot \mathbf{t} + \mathbf{n} \cdot \mathbf{t}' = 0$$
$$\Rightarrow \; \mathbf{n}' \cdot \mathbf{t} = -\mathbf{n} \cdot \mathbf{t}' = -\mathbf{n} \cdot k\mathbf{n} = -k$$
$$\Rightarrow \; \gamma = -k.$$

Therefore

$$\mathbf{n}' = \tau\mathbf{b} - k\mathbf{t}.$$

Now if $\mathbf{q}(s)$ is a regular smooth curve parametrized by arc léngth then:

(i) If $k(s) = 0$ for all real values of s then the curve is a straight line.
(ii) If $\tau(s) = 0$ for all real values of s then the curve lies in a plane.

Proof

(i) $k = 0$

$$\Rightarrow \; \mathbf{q}'' \; = \; 0$$
$$\Rightarrow \; \mathbf{q}' \; = \; \mathbf{c}$$
$$\Rightarrow \; \mathbf{q}(s) \; = \; s\mathbf{c} + \mathbf{d}$$

which is the equation of a plane for some vector constants $\mathbf{c}$, $\mathbf{d}$.

(ii) $\tau = 0$

If $\tau = 0$ then Eq. 30 implies that $\mathbf{b}' = 0$ and so $\mathbf{b} =$const. Thus, the plane defined by t and n (to which b is perpendicular) remains constant, i.e. the curve lies in a plane (Fig. 12).

Example

Consider the circular helix parametrized by arc length:

$$\mathbf{q}(s) = \left(a \cos \frac{s}{\sqrt{a^2 + b^2}}, a \sin \frac{s}{\sqrt{a^2 + b^2}}, \frac{bs}{\sqrt{a^2 + b^2}} \right).$$

Then

$$\mathbf{t} = \mathbf{q}'(s) = \left(\frac{-a}{\sqrt{a^2 + b^2}} \sin \frac{s}{\sqrt{a^2 + b^2}}, \frac{a}{\sqrt{a^2 + b^2}} \cos \frac{s}{\sqrt{a^2 + b^2}}, \frac{b}{\sqrt{a^2 + b^2}} \right)$$

$$\Rightarrow \quad \mathbf{t}' = \mathbf{q}''(s) = \left(\frac{-a}{(a^2 + b^2)} \cos \frac{s}{\sqrt{a^2 + b^2}}, \frac{-a}{(a^2 + b^2)} \sin \frac{s}{\sqrt{a^2 + b^2}}, 0 \right).$$

The curvature
$$k = |\mathbf{t}'| = \frac{a}{(a^2 + b^2)}$$

$$\mathbf{n} = \frac{\mathbf{t}'}{k} = \left(-\cos \frac{s}{\sqrt{a^2 + b^2}}, -\sin \frac{s}{\sqrt{a^2 + b^2}}, 0 \right)$$

$$\Rightarrow \quad \mathbf{b} = \mathbf{t} \times \mathbf{n} = \left(\frac{b}{\sqrt{a^2 + b^2}} \sin \frac{s}{\sqrt{a^2 + b^2}}, \frac{-b}{\sqrt{a^2 + b^2}} \cos \frac{s}{\sqrt{a^2 + b^2}}, \frac{b}{\sqrt{a^2 + b^2}} \right)$$

Therefore
$$\mathbf{b}' = \left(\frac{b}{(a^2 + b^2)} \cos \frac{s}{\sqrt{a^2 + b^2}}, \frac{b}{(a^2 + b^2)} \sin \frac{s}{\sqrt{a^2 + b^2}}, 0 \right)$$

$$= \frac{-b}{(a^2 + b^2)} \cdot \mathbf{n}$$

$$\Rightarrow \quad \tau = \frac{b}{(a^2 + b^2)}.$$

Hence k, τ are constants and $\tau/k = b/a$.

Therefore the helix has the normal $\mathbf{n}$ always pointing towards the axis of the cylinder and $k(s)$ is constant due to the circular motion of the helix. Also if $\mathbf{b} = \mathbf{0}$ then $\tau = 0$ and $\mathbf{q}(s)$ represents a circle which is a curve with constant curvature.

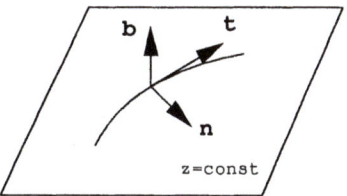

Fig. 12. The curve in the plane $z = const$

4.1. Alternative Form of the Serret Frenet Formulae

Suppose we have a curve $\mathbf{r}(t)$ not parametrized by arc length (which is the more usual situation in CAGD). Then we can still evaluate the curvature and torsion of these curves, without having to find the arc length function s. We will represent differentiation with respect to arc length s by a dash and differentiation with respect to the parameter t by a dot.

Using the chain rule for differentiation

$$\dot{\mathbf{r}} = \frac{d\mathbf{r}}{dt} = \frac{ds}{dt} \cdot \frac{d\mathbf{r}}{ds} = \dot{s}\mathbf{r}' \tag{32}$$

for some parametrization t.

From Eq. 15 for the definition of s we obtain

$$\dot{s} = |\dot{\mathbf{r}}|. \tag{33}$$

Now

$$
\begin{aligned}
\dot{\mathbf{r}} &= \dot{s}\mathbf{r}' = \dot{s}\mathbf{t} \\
\Rightarrow \ddot{\mathbf{r}} &= \ddot{s}\mathbf{t} + \dot{s}\dot{\mathbf{t}} \\
\text{but } \dot{\mathbf{t}} &= \dot{s}\mathbf{t}' = \dot{s}k\mathbf{n} \\
\Rightarrow \ddot{\mathbf{r}} &= \ddot{s}\mathbf{t} + \dot{s}^2 k\mathbf{n} \\
\Rightarrow \dot{\mathbf{r}} \times \ddot{\mathbf{r}} &= \dot{s}^3 k(\mathbf{t} \times \mathbf{n}) = \dot{s}^3 k\mathbf{b}.
\end{aligned}
$$

Since $\dot{s} > 0$ we thus get an expression for the curvature given by

$$k = \frac{|\dot{\mathbf{r}} \times \ddot{\mathbf{r}}|}{\dot{s}^3} = \frac{|\dot{\mathbf{r}} \times \ddot{\mathbf{r}}|}{|\dot{\mathbf{r}}|^3}. \tag{34}$$

Following a similar procedure

$$
\begin{aligned}
\dddot{\mathbf{r}} &= \dddot{s}\,\mathbf{t} + \ddot{s}\dot{\mathbf{t}} + 2\dot{s}\ddot{s}k\mathbf{n} + \dot{s}^2\dot{k}\mathbf{n} + \dot{s}^2 k\dot{\mathbf{n}} \\
\Rightarrow (\dot{\mathbf{r}} \times \ddot{\mathbf{r}}) \cdot \dddot{\mathbf{r}} &= \dot{s}^3 k\ddot{s}(\mathbf{b} \cdot \dot{\mathbf{t}}) + \dot{s}^3 k\dot{s}^2(k\mathbf{b} \cdot \dot{\mathbf{n}}).
\end{aligned}
$$

But, $\dot{\mathbf{t}} = \dot{s}k\mathbf{n}$ and $\dot{\mathbf{n}} = \dot{s}\mathbf{n}'$

$$\Rightarrow (\dot{\mathbf{r}} \times \ddot{\mathbf{r}}) \cdot \dddot{\mathbf{r}} = \dot{s}^6 k^2 \tau$$

So from these we have expressions to find the curve properties from a parametrization given by

$$k = \frac{|\dot{\mathbf{r}} \times \ddot{\mathbf{r}}|}{|\dot{\mathbf{r}}|^3}, \qquad \tau = \frac{(\dot{\mathbf{r}} \times \ddot{\mathbf{r}}) \cdot \dddot{\mathbf{r}}}{|\dot{\mathbf{r}} \times \ddot{\mathbf{r}}|^2} \tag{35}$$

and

$$\mathbf{t} = \frac{\dot{\mathbf{r}}}{|\dot{\mathbf{r}}|}, \qquad \mathbf{b} = \frac{\dot{\mathbf{r}} \times \ddot{\mathbf{r}}}{|\dot{\mathbf{r}} \times \ddot{\mathbf{r}}|} \tag{36}$$

and finally

$$\mathbf{n} = \mathbf{b} \times \mathbf{t}. \tag{37}$$

4.2. The Frenet Frame

In the neighourhood of the point s of the curve $\mathbf{q}(s)$ we have a natural coordinate system, the Frenet Frame [73] at s given by $\mathbf{t}$, $\mathbf{n}$ and $\mathbf{b}$.

Within this framework we define the three planes:

o The osculating plane spanned by **t**, **n**.

o The normal plane spanned by **n**, **b**.

o The rectifying plane spanned by **b**, **t**.

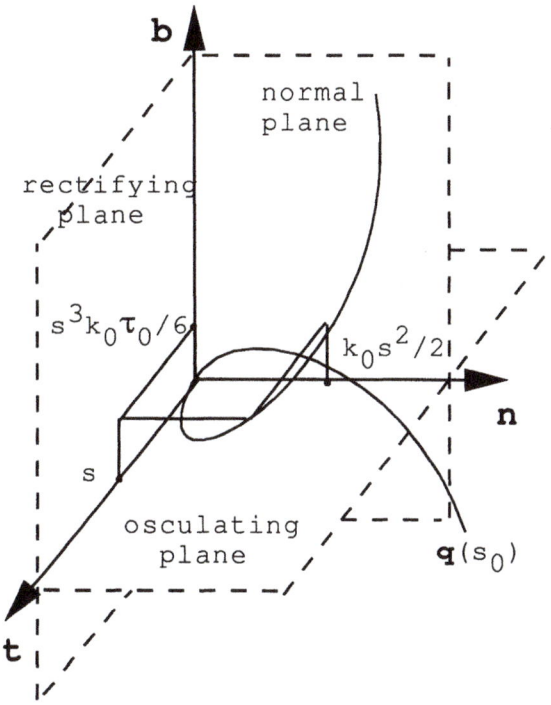

Fig. 13. The Frenet frame

Therefore at the point $\mathbf{q}(s_0)$ we can define the three planes by

$$
\begin{array}{llr}
\text{Osculating plane} & (\mathbf{r} - \mathbf{q}(s_0)) \cdot \mathbf{b} = 0 & (38) \\
\text{Rectifying plane} & (\mathbf{r} - \mathbf{q}(s_0)) \cdot \mathbf{n} = 0 & (39) \\
\text{Normal plane} & (\mathbf{r} - \mathbf{q}(s_0)) \cdot \mathbf{t} = 0 & (40)
\end{array}
$$

If we now consider a unit speed curve in the vicinity of the point $\mathbf{q}(s_0)$ we can expand this as a Taylor series about $s = s_0$ and without loss of generality assume that $s_0 = 0$, thus

$$
\mathbf{q}(s) = \mathbf{q}(s_0) + s\mathbf{q}'(s_0) + \frac{s^2}{2}\mathbf{q}''(s_0) + \frac{s^3}{6}\mathbf{q}'''(s_0) + R \tag{41}
$$

where

$$\lim_{s \to 0} \frac{R}{s^3} = 0. \tag{42}$$

But remembering that $\mathbf{q}'(0) = \mathbf{t}$, $\mathbf{q}''(0) = k\mathbf{n}$

$$\Rightarrow \mathbf{q}'''(0) = (k\mathbf{n})' = k'\mathbf{n} + k\mathbf{n}' = k'\mathbf{n} - k^2\mathbf{t} + k\tau\mathbf{b}$$

$$\Rightarrow \mathbf{q}(s) - \mathbf{q}(0) = s\mathbf{t_0} + \frac{s^2 k_0}{2}\mathbf{n_0} + \frac{s^3}{6}k_0\tau_0\mathbf{b_0} \tag{43}$$

where we are retaining only the dominant terms in each direction. This is the Frenet approximation to the curve. The term involving $\mathbf{t_0}$ gives the best linear approximation to the curve, while the terms involving $\mathbf{t_0}$ and $\mathbf{n_0}$ give the best quadratic approximation to the curve. The term parallel to $\mathbf{b_0}$ controls the movement of the curve away from the osculating plane, all of which can be seen in Fig. 13.

So if x, y, z are coordinates of $\mathbf{q}(s)$ taken along $\mathbf{t_0}$, $\mathbf{b_0}$ and $\mathbf{n_0}$ we have

$$x = s + \cdots \tag{44}$$

$$y = \frac{s^2 k_0}{2} + \cdots \tag{45}$$

$$z = \frac{s^3}{6}k_0\tau_0 + \cdots \tag{46}$$

and if we project the curve onto each of the three planes we obtain the following curves shown overleaf in Fig. 14.

1. $z = \frac{k_0 \tau_0}{6}x^3$ $(\mathbf{q}(s) = s\mathbf{t_0} + \frac{s^3}{6}k_0\tau_0\mathbf{b_0})$ the rectifying plane

2. $y = \frac{k_0}{2}x^2$ $(\mathbf{q}(s) = s\mathbf{t_0} + \frac{s^2}{2}k_0\mathbf{n_0})$ the osculating plane

3. $2\tau_0^2 y^3 = 9k_0 z^2$ $(\mathbf{q}(s) = \frac{s^2}{2}k_0\mathbf{n_0} + \frac{s^3}{6}k_0\tau_0\mathbf{b_0})$ the normal plane

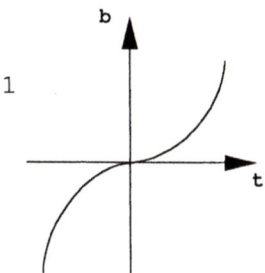

Projection onto the rectifying plane

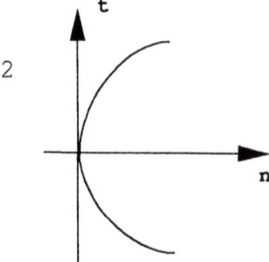

Projection onto the osculating plane

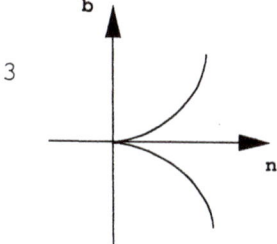

Projection onto the normal plane

Fig. 14. The projection of the curve onto the Frenet frame

CHAPTER 3
CURVE REPRESENTATION

1. Analytic Representation of a Curve

Geometric modelling of curves is one of the fundamental problems in CAGD. It is usually the starting point for the modelling of surfaces or more complex objects which is the essence of CAGD.

Geometric modelling of a curve is an iterative and usually interactive process. It leads to conceptualization of a curve shape which gradually refines ideas going from an initial image (often through a hand sketch) to the final, acceptable form. There are two principal stages in this process with different objectives and rules:

- concept design stage:
    definition and generation of a curve,

- final design stage:
    fairing of a curve [50].

In this chapter the interest will be turned to the concept design stage.

Geometric modelling of curves needs an appropriate analytic curve representation. The general aspects of such a representation have been discussed in Chapter 2. It was said that in CAGD an arbitrary curve, being a geometric object (a collection of points in 3-dimensional Euclidean space E^3), can be regarded as the mapping of a vector-valued function $\mathbf{r}(t)$ of the type: $\mathbf{r} : R^1 \to E^3$. In practice, the curve itself is identified with its function $\mathbf{r}(t)$ which can be expressed in the following matrix form:

$$\mathbf{r}(t) = \begin{bmatrix} r_1(t) \\ r_2(t) \\ r_3(t) \end{bmatrix} = \begin{bmatrix} x(t) \\ y(t) \\ z(t) \end{bmatrix} = \sum_{i=1}^{3} r_i(t) \cdot \mathbf{e}_i \tag{47}$$

where

$\mathbf{e}_i$ - basis of the orthogonal coordinate system $(0, x, y, z)$,

$r_i : R^1 \to R^1$ - components of $\mathbf{r}$ (single-valued functions corresponding to the axes x, y, z),

$t \in [a, b]-$ domain of the function $\mathbf{r}$.

Such a form is called a parametric curve representation. For one specific value of t_i there exists only one point P_i on the curve $P_i = \mathbf{r}(t_i)$, but conversely, one specific point on the curve may corresponds to many values of t. This mutual association of points and parameters is called parametrization of a curve.

In order to define $\mathbf{r}(t)$ analytically, for practical purposes of geometric modelling, a general structure of the function $\mathbf{r}(t)$ and a specific class of the functions $r_i(t)$ has to be chosen. The simplest possible structure of any function is a linear structure.

1.1. Algebraic Form of Curve Definition

The linear structure of the function $r(t)$ can be written most easily in matrix notation. The three equivalent forms that may be used are:

$$\mathbf{r}(t) = \begin{bmatrix} r_1(t) \\ --- \\ r_2(t) \\ --- \\ r_3(t) \end{bmatrix} = \begin{bmatrix} \sum_{j=0}^{n} a_{1j} \cdot E_j(t) \\ \sum_{j=0}^{n} a_{2j} \cdot E_j(t) \\ \sum_{j=0}^{n} a_{3j} \cdot E_j(t) \end{bmatrix} = \begin{cases} (\hat{\mathbf{A}}_1, \hat{\mathbf{A}}_2, \hat{\mathbf{A}}_3)^T \cdot \mathbf{E}(t) \\ \text{or} \\ \mathbf{A} \cdot \mathbf{E}(t) \\ \text{or} \\ (\mathbf{A}_0, \mathbf{A}_1, \cdots, \mathbf{A}_n) \cdot \mathbf{E}(t) \end{cases} \tag{48}$$

where $\mathbf{E}(t) = (E_1(t), E_2(t), \cdots, E_n(t))^T$ is a vector of functions called basis functions, and

$$\mathbf{A} = \begin{bmatrix} a_{10} & | & a_{11} & | & a_{12} & | & \cdots a_{1n} \\ --- & | & --- & | & --- & | & --- \\ a_{20} & | & a_{21} & | & a_{22} & | & \cdots a_{2n} \\ --- & | & --- & | & --- & | & --- \\ a_{30} & | & a_{31} & | & a_{32} & | & \cdots a_{3n} \end{bmatrix} \begin{matrix} \equiv \hat{\mathbf{A}}_1^T \\ \\ \equiv \hat{\mathbf{A}}_2^T \\ \\ \equiv \hat{\mathbf{A}}_3^T \end{matrix} \quad - \text{ matrix of coefficients.}$$

$$\mathbf{A}_0 \qquad \mathbf{A}_1 \qquad \mathbf{A}_2 \qquad \cdots \mathbf{A}_n \tag{49}$$

The symbol ^{T} used in the definitions above, is the sign of matrix transposition, and the symbol $\cdot$ means matrix multiplication. In the chapters that follow, all vectors are described by single-column matrices.

The basis functions $E_j(t)$ constitute a functional generalization of basis vectors like those in Eq. 47. In contrast to the real space E^3, it spans an abstractive, $(n+1) = N$ - dimensional space of functions X_N, so $\mathbf{r} \in X_N$. In order for the functions E_j to be a basis of X_N they must be linearly independent. Note, that by assumption, the basis $\mathbf{E}(\cdot)$ in the definition (Eq. 48) is the same for all components $r_i(t)$.

The matrix $\mathbf{A}$ (Eq. 49) contains all the coefficients a_{ij} which constitute 'degrees of freedom' of the function $\mathbf{r}$ and are to be determined in the geometric modelling process. For convenience, the matrix $\mathbf{A}$ can be expressed as a composition of vectors, $\hat{\mathbf{A}}_1, \cdots$, or $\mathbf{A}_0, \cdots$, alternatively (Eq. 48). The parametric curve representation in the form of Eq. 48 is sometimes called an algebraic form of curve representation [47].

The linear structure (Eq. 48) has many advantages. Among them, it eases all linear operations (like those of differentiation, integration, etc.) performed on a function $\mathbf{r}$. Let $L(\cdot)$ be an arbitrary linear operator acting on $\mathbf{r}$. Then, by virtue of linearity one gets

$$L(\mathbf{r})(t) = L(\mathbf{A} \cdot \mathbf{E}(t)) = \mathbf{A} \cdot L(\mathbf{E})(t). \tag{50}$$

For example, let $L(\cdot)$ be the operator of differentiation:

$$L(\mathbf{r})(t) = \frac{d^k}{dt^k} \mathbf{r}(t) \equiv \mathbf{r}^k(t), \quad k = 0, 1, \cdots \tag{51}$$

Then,

$$\mathbf{r}^k(t) = \mathbf{A} \cdot \mathbf{E}^k(t). \tag{52}$$

This reduces a linear operation on a function to the linear operation on a basis and makes it possible to use a compact matrix notation.

1.2. Parametrization of a Curve

Parametrization of a curve is a procedure which associates the points on the curve $\mathbf{P}_i$ with the specific parameter values t_i. This problem does not arise when the curve is described by a scalar function (e.g. $y = f(x)$). The issue of how to specify these parameters, that is, how to 'parametrize' a curve is neither trivial nor unique. Any choice made will have an influence on the shape of the curve.

In mathematical terms, a parametrization P of a curve means a mapping 'points onto numbers', so, in general, $P : E^3 \rightarrow R^1$. In particular, however, the mapping can be defined as a non-decreasing sequence (special kind of single-valued function where arguments are integers) : $P : i \rightarrow t_i$, where i are indices of successive points on a curve (Fig. 15).

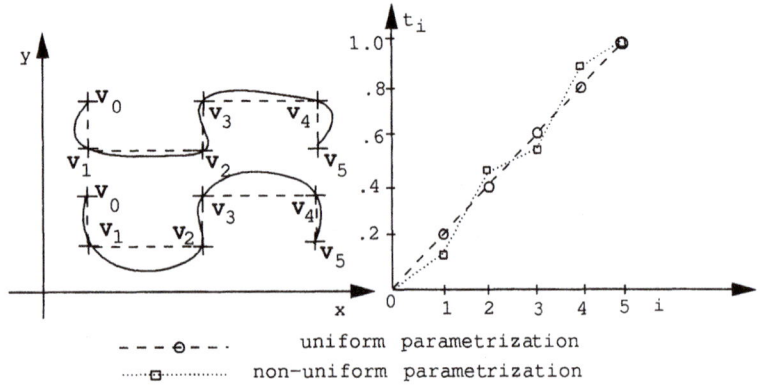

Fig. 15. Parametrization of a curve

If the function t_i is a linear function (t_i is an arithmetic progression) then the parametrization of the curve is called a uniform parametrization, otherwise it is called a non-uniform parametrization.

The simplest examples of uniform parametrization of a curve (being standard parametrizations in CAGD) are

$$P : t_i = i, \quad i = 0, 1, 2, \cdots, n \ \Rightarrow t \in [0, n] \tag{53}$$

and

$$P : t_i = \frac{i}{n}, \quad i = 0, 1, 2, \cdots, n \ \Rightarrow t \in [0, 1]. \tag{54}$$

The procedure of changing one parametrization of a curve to another is called reparametrization of the curve. If a new parameter of the curve is taken as u, then reparametrization defines a function: $u = u(t)$ (see Chapter 2).

The necessity of reparametrization appears when one wants to normalize the domain of a curve (Eq. 54), to make a partition of a curve, to join another segment of a curve, or to introduce a non-uniform parametrization of a curve, etc. While modelling a curve in CAGD, the uniform parametrization is usually an initial choice. Then, if necessary, the non-uniform parametrization is applied, which offers some additional 'degrees of freedom' and helps to modify the shape of the curve.

The basic non-uniform parametrization of a curve is a natural parametrization, in which successive values of t_i are equal to arc lengths $\mathbf{P}_0\mathbf{P}_i$, $i = 0, 1, 2, \cdots, n$ along the curve (see Chapter 2). The natural parametrization is of great importance in the theory of curves, but has rather limited application in CAGD.

The standard non-uniform parametrization of a curve in CAGD is the accumulated chord length parametrization, according to the definition

$$u(t_i) = \begin{cases} 0 & \text{for } i = 0 \\ \sum_{j=1}^i \| \mathbf{P}(t_j) - \mathbf{P}(t_{j-1}) \| & \text{for } i = 1, 2, \cdots, n \end{cases} \tag{55}$$

or, after normalization

$$\overline{u}(t_i) = \frac{u(t_i)}{\sum_{j=1}^n \| \mathbf{P}(t_j) - \mathbf{P}(t_{j-1}) \|}; \quad \overline{u} \in [0, 1]. \tag{56}$$

Therefore, the accumulated chord length parametrization approximates the natural parametrization of a curve.

The parameters of a curve t_j corresponding to its characteristic points $\mathbf{P}_j$ are called knots. A non-decreasing sequence of knots i.e. $t_j \le t_{j+1}$ can be expressed as a vector of knots: $\mathbf{T} = (t_0, t_1, \cdots, t_n)$. If $t \in [a, b]$ then the vector of knots $\mathbf{T}$ determines a partition of the interval $[a, b]$ into subintervals $[t_i, t_{i+1}]$, $i = 0, 1, 2, \cdots, n - 1$, π: $a = t_0 \le t_1 \le t_2 \le \cdots \le t_n = b$.

A curve for which $\mathbf{P}(t_0) \equiv \mathbf{P}(t_n)$ is called a closed (or periodic) curve, otherwise it is called open (or non-periodic).

1.3. Parametric Polynomial Curves

The class of a function $\mathbf{r}(t)$ possessing the linear structure (Eq. 48) is completely determined by the basis $\mathbf{E}(t)$. The most common basis used in CAGD is the following power basis:

$$\mathbf{E}_n(t) = [1, t, t^2, t^3, \cdots, t^n]^T. \tag{57}$$

It makes all the functions $r_i(t)$ in Eq. 48 ordinary polynomials and the corresponding curve $\mathbf{r}(t)$ a parametric polynomial curve of degree $p = n$ and order $k = n + 1$. According to Eq. 50 it can be expressed in the matrix form:

$$\mathbf{r}(t) = [1, t, t^2, t^3, \cdots, t^n] \cdot \mathbf{A}^T = \mathbf{A} \cdot \mathbf{E}_n(t). \tag{58}$$

The lowest degree n to describe an arbitrary space curve is $n = 3$ (cubic curve). For $n < 3$ Eq. 48 with the basis given by Eq. 57 describes only planar curves (situated in space arbitrarily).

The polynomial curve representation (Eq. 51) can be regarded as a fundamental (in a sense) mathematical representation of a curve in CAGD despite the fact that the basis $\mathbf{E}(\cdot)$ (sometimes called the Weierstrass' basis) is not the only basis for polynomial curve representation. In fact, it is never used in CAGD directly, for many reasons.

The parametric polynomial curve is an elementary function and as such is infinite times differentiable (the derivatives of the degree $p > n$ are identically equal to zero). So, in the case of Eq. 51: $\mathbf{r} \in C^\infty$. This property of the polynomial curve has an advantage in analytical considerations. In CAGD however, on the contrary, it has appeared as a great disadvantage of ordinary polynomial curves. The infinite differentiability of the curve has turned out to be neither useful nor necessary. Abandoning infinite differentiability in favour of finite differentiability in a finite number of points has led to the concept of parametric piecewise polynomials and splines.

1.4. Parametric Piecewise and Spline Curves

The idea of parametric piecewise polynomial curves and parametric polynomial spline curves was to define a polynomial curve not globally, on the whole domain of the curve, but locally, on the successive intervals of the domain and then to join separate segments together to form one curve.

Let $\pi : a = t_0 \leq t_1 \leq t_2 \leq \cdots \leq t_n = b$, be a partition of $[a, b]$ into sub-intervals $[t_i, t_{i+1}]$. Then a function $\mathbf{r}(t)$ $t \in [a, b]$ is called a parametric piecewise polynomial function of degree $p \leq n$ (order $k = p + 1$) if $\mathbf{r}(t)$ is a parametric polynomial function (Eq. 58) of degree p, $p = 1, 2, 3, \cdots$ on each subinterval $[t_i, t_{i+1}]$, $i = 0, 1, \cdots, n-1$.

In such a way one can obtain piecewise polynomial curves. These depend on the degree p. They are linear for $p = 1$, quadratic ($p = 2$), cubic ($p = 3$), quartic ($p = 4$), quintic ($p = 5$) and so on. It follows from the definition that in each of these cases $\mathbf{r}(t)$ may fail to be differentiable at some or all of its knots t_i. This is because the conditions of smoothness at knots are not specified in the definition. The case in which these conditions are specified leads to the concept of spline functions.

A function $\mathbf{r}(t)$, $t \in [a, b]$ is called a parametric polynomial spline function of degree $p = n$ (order $k = p + 1$) if $\mathbf{r}(t)$ is a parametric piecewise polynomial function of degree p : $p = 1, 2, 3, \cdots$, on each subinterval $[t_i, t_{i+1}]$, $i = 0, 1, \cdots, n-1$ and $\mathbf{r}(t)$ and its derivatives of orders $0, 1, \cdots, p - 1$ are everywhere continuous, that is $\mathbf{r} \in C^{p-1}$.

It can be observed, that for $p = n$ both piecewise polynomials and polynomial splines become ordinary polynomial functions. This means that a polynomial function is a special case of spline function.

A spline function is a mathematical model of a physical spline - the drafting tool, familiar in several industries (including the shipbuilding industry). The physical spline (Fig. 16) is a thin, elastic lath, usually made of wood, metal or plastic and used to draw smooth curves through given points. The lath is held fixed in certain places by heavy weights called ducks which constrain the curve to run through these points. The type of support may be regarded as hinged. Then the system of spline and ducks may be compared to a simply-supported, thin, continuous elastic beam. Assuming small deflections and hence using linearized beam theory one can show:

- The deflection in each span can be represented by a cubic polynomial.

- The bending moments, which are proportional to the second derivatives of the deflection, are continuous through the entire length of the spline, that is, the curve is C^2 continuous.

- The third derivatives are discontinuous at the supports unless the supporting force happens to be zero.

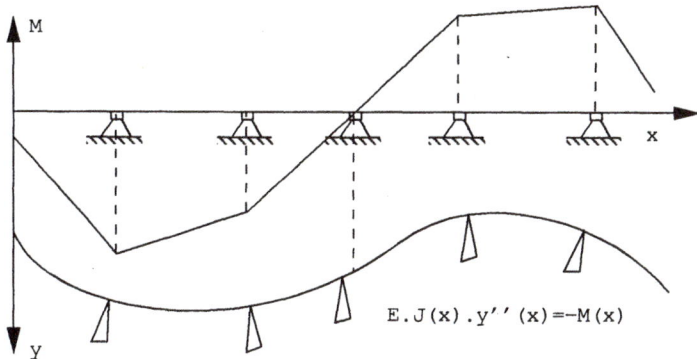

Fig. 16. A physical spline and its mechanical model

It is evident that this behaviour of the drafting tool can readily be simulated mathematically by a cubic spline ($p = 3, k = 4$) which has exactly the same three properties. This is one reason why cubic spline functions have found wide-spread application in CAGD.

It can be shown that different types of parametric piecewise polynomial functions and parametric polynomial spline functions have their own bases. However, all of them can also be expressed in the fundamental form (Eq. 58) by an ordinary power basis.

The advantages of parametric piecewise polynomial functions and, in particular, parametric polynomial spline functions become visible when solving approximation and interpolation problems which are basic problems in the geometric modelling of curves in CAGD [34].

1.5. Approximation and Interpolation Problems

The problem of geometric modelling of a curve can be stated as follows [50]:

" Given a set of geometric requirements such as: offsets, slopes, second derivatives, areas, etc., find an acceptable curve r(t) meeting these requirements."

In mathematical terms this problem generally corresponds to an approximation problem. One function, $\mathbf{f}(t)$, is approximated by another one, $\mathbf{r}(t)$, such that a measure of distance between them $\| \mathbf{f} - \mathbf{r} \|$ will be small, subject to a set of constraints. In general, the curve $\mathbf{f}(t)$ is given by its analytical definition like that of Eq. 47 (continuous manner). In particular, however, it may be given solely by a number of points on it (discrete manner). The latter case is typical in CAGD.

The type of constraints determines the type of approximation problem [64]:

a. Pure interpolatory constraints, e.g:
$$\mathbf{r}(t_i) = \mathbf{f}(t_i), \quad i = 0, 1, \cdots, n \quad t_i \in [a, b]$$

b. Mixed interpolatory constraints, e.g:
$$\mathbf{r}^{k_i}(t_i) = \mathbf{f}^{k_i}(t_i), \quad i = 0, 1, \cdots, n \quad k_i = 0, 1, 2, \cdots, p$$

c. Pure variational constraints, e.g:
$$\| \mathbf{f} - \mathbf{r} \| = \min \{ \| \mathbf{f} - \mathbf{r} \| : \mathbf{r} \in \mathbf{X}_n - n\text{-dimensional space} \}$$

d. Mixed variational / interpolatory constraints, etc.

$$(59)$$

These analytical constraints fully correspond to those of the geometric requirements mentioned above. If the requirements are treated as 'hard constraints' so that the curve must meet them exactly (e.g. pass through a number of points), then the problem is briefly referred to as an interpolating problem. If, by contrast, a violation of the requirements is feasible and the constraints are acting only as 'soft constraints' the problem is briefly referred to as an approximation problem. In practice, both of these actually fall in category '59d' of constraints. Fig. 17 illustrates both problems.

The norm $\| \mathbf{f} - \mathbf{r} \|$ is, in any case, a 'measure of merit' of the approximation. However, in CAGD it is neither minimized nor even calculated directly. Instead, it is taken into account indirectly by a proper choice of $\mathbf{r}$. Polynomial spline functions are those which have excellent properties both in pure interpolation as well as in pure approximation solutions.

1.6. Geometric Form of Curve Definition

The algebraic form of a parametric curve definition (Eq. 48), being the most general form, is not convenient in practical applications in CAGD, especially when

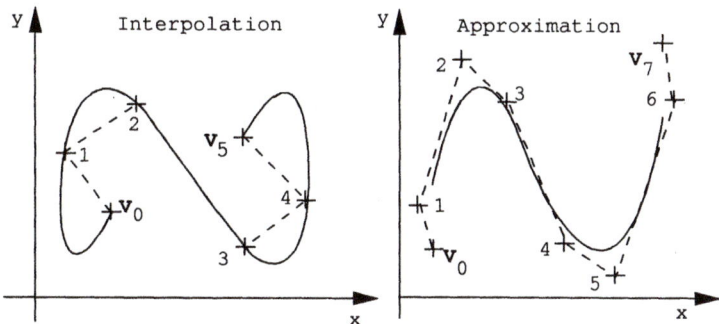

Fig. 17. Interpolation and approximation problems

modelling the curve interactively. This is mainly because the algebraic coefficients a_{ij}, expressed as a matrix $\mathbf{A}$, have no clear geometric interpretations and so they do not contribute much to an intuitive meaning of the curve. By contrast, the geometric form of a curve definition seems to fulfill these needs.

Let us consider a mixed interpolatory problem and define it by linear operators as in Eq. 50. The problem then reduces to the following system of linear equations:

$$L(\mathbf{r})(t_m) = \mathbf{Q}_m \Leftrightarrow \mathbf{A} \cdot L(\mathbf{E})(t_m) = \mathbf{Q}_m, \quad m = 0, 1, 2, \cdots, n. \tag{60}$$

where $\mathbf{Q}_m = (Q_{m1}, Q_{m2}, Q_{m3})^T$ are vectors of geometric requirements ('parameters') imposed on the curve under design. The system (Eq. 60) can be expressed in matrix form as a single matrix equation:

$$\mathbf{A} \cdot \mathbf{B} = \mathbf{Q} \tag{61}$$

where $\mathbf{B} = [b_{ij}] = [L(E_i)(t_j)]$ and $\mathbf{Q} = [Q_0, Q_1, \cdots, Q_n]$, for $i, j = 0, 1, 2, \cdots, n$. If $\det \mathbf{B} \neq 0$ (the solution of the interpolation does exist, which is a separate problem), then Eq. 61 can be solved with respect to $\mathbf{A}$:

$$\mathbf{A} = \mathbf{Q} \cdot \mathbf{B}^{-1}. \tag{62}$$

Substituting Eq. 62 into Eq. 48 yields

$$\mathbf{r}(t) = \mathbf{A} \cdot \mathbf{E}(t) = \mathbf{Q} \cdot \mathbf{B}^{-1} \cdot \mathbf{E}(t) = \mathbf{Q} \cdot \mathbf{F}(t) \tag{63}$$

and finally, one gets two equivalent forms of $\mathbf{r}(t)$:

$$\mathbf{r}(t) = [\mathbf{A}_0, \mathbf{A}_1, \cdots, \mathbf{A}_n] \cdot \mathbf{E}(t) = \mathbf{A} \cdot \mathbf{E}(t), \tag{64}$$
$$\mathbf{r}(t) = [\mathbf{Q}_0, \mathbf{Q}_1, \cdots, \mathbf{Q}_n] \cdot \mathbf{F}(t) = \mathbf{Q} \cdot \mathbf{F}(t). \tag{65}$$

Eq. 64 is called an algebraic form and Eq. 65 is called a geometric form of parametric curve definition [47]. This is because in this form the parameters $\mathbf{Q}_m$

(coefficients of the curve equation) have a clear geometric interpretation determined by the linear operators (Eq. 60). One can prove that if $\mathbf{E}(t)$ is a basis of $\mathbf{X}_N$ then $\mathbf{F}(t)$ is also a basis of $\mathbf{X}_N$. In the case of interpolation, the basis $\mathbf{F}(t)$ is called a cardinal basis [3],[64]. If one knows the cardinal basis for a specific interpolation problem one does not need to solve a system of equations (Eq. 60), as Eq. 65 is defined.

It will be shown that the geometric form of parametric curve definition applies to approximation problems as well. In general, for both problems, the basis functions $F_j(t)$ of geometric form definition (Eq. 65) are called blending functions.

In CAGD all geometric modelling methods are based on interpolating or approximating techniques. All of them make use of the geometric form of curve definition.

2. Interpolation Techniques

It follows from Section 1.5 that an interpolation problem is a special case of an approximation problem. Depending on the type of (interpolatory) constraints (Eq. 59a,b) it can be solved by different techniques. The main techniques used in CAGD are two classical techniques:

- Lagrange's technique (J. L. Lagrange, 1763-1813),

- Hermite's technique (Ch. Hermite, 1822-1901),

and a special case of Hermite's technique:

- Ferguson's technique,

plus other modern spline techniques, introduced to CAGD by Schoenberg in 1946.

In all the interpolation techniques a curve (function) to be interpolated is given in a discrete manner, at least by a sequence of pairs $(t_i, \mathbf{P}_i)$, $i = 0, 1, 2, \cdots, n$, where t_i are knots and $\mathbf{P}_i$ are successive points on the curve. The points $\mathbf{P}_i$ are called control points or alternatively position vectors. In some interpolation techniques this information about the curve is additionally complemented by tangent vectors $\mathbf{Q}_i$. Both position vectors and tangent vectors are also called the control vectors. Fig. 18 illustrates the idea for a planar curve $\mathbf{r}(t)$.

All neighbouring position vectors connected to each other by straight lines form a defining polygon $P_n = (\mathbf{V}_0, \mathbf{V}_1, \cdots, \mathbf{V}_n)$ with vertices $\mathbf{V}_i$ and edges $\overline{\mathbf{V}_i\mathbf{V}_{i+1}}$, $i = 0, 1, \cdots, n - 1$. In all the interpolating techniques the curve to be found, $\mathbf{r}(t)$, is forced to pass exactly through the vertices of P_n so in this case $\mathbf{V}_i = \mathbf{P}_i$. The defining polygon P_n can be regarded as a piecewise linear function $\mathbf{f}(t)$ (Eq. 59) to be approximated.

The result of interpolation $\| \mathbf{f} - \mathbf{r} \|$ highly depends on the desirable smoothness s of $\mathbf{r}(t)$. For $s = 0$ (positional continuity only) the polygon P_n itself is the solution of the problem. In CAGD, however, higher degrees of smoothness ($s = 1, 2$) are usually expected. It is known [64] that the higher the degree of smoothness desired, the bigger the norm $\| \mathbf{f} - \mathbf{r} \|$ is, so the worse the interpolation obtainable. This suggests a good compromise.

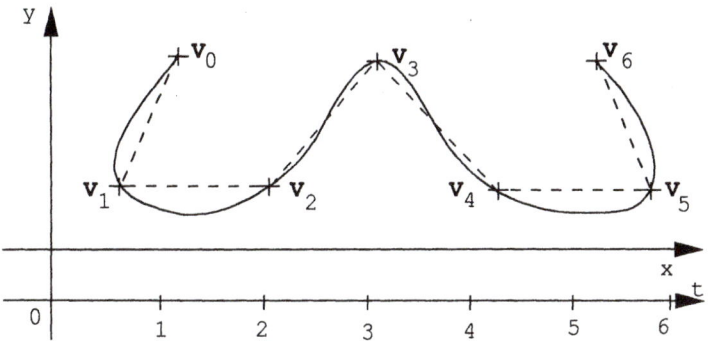

Fig. 18. Interpolation technique - control vectors

2.1. Lagrange Interpolation

This is the simplest case of interpolation where the pure interpolatory constraints (Eq. 59a) are applied, so the curve to be found $r(t)$ is determined by a vector of knots $\mathbf{T} = (t_0, t_1, \cdots, t_n)$ and corresponding sequence of position vectors $\mathbf{P}_i = \mathbf{V}_i$, $i = 0, 1, \cdots, n$.

In mathematical terms, the Lagrange interpolation leads to the following system of equations:

$$r(t_i) = \mathbf{P}_i \quad i = 0, 1, \cdots, n. \tag{66}$$

If, by assumption, the curve $r(t)$ is the polynomial parametric curve, the problem is called Lagrange polynomial interpolation. In such a case the unique solution of Eq. 66 exists [64] when $r(t)$ is a parametric polynomial curve of degree $p = n$.

The solution of Lagrange interpolation can instantly be obtained when the geometric form (Eq. 65) of the curve $r(t)$ is applied:

$$r(t) = [\mathbf{P}_0, \mathbf{P}_1, \cdots, \mathbf{P}_n] \cdot \mathbf{F}(t) = \mathbf{P} \cdot \mathbf{F}(t). \tag{67}$$

The cardinal basis of the Lagrange interpolation: $L_i(t) \equiv F_i(t)$, for $t \in [0, 1]$, is defined as follows [47],[83]:

$$L_i(t) = \frac{w_i(t)}{w_i(t_i)} = \frac{w(t)}{(t - t_i) \cdot w_i(t_i)} = \frac{w(t)}{(t - t_i) \cdot w'(t_i)} = \prod_{j=0, j \neq i}^{n} \frac{(t - t_j)}{(t_i - t_j)} \tag{68}$$

where

$$w(t) = (t - t_0)(t - t_1) \cdots \cdots (t - t_n) \equiv \prod_{j=0}^{n} (t - t_j) \tag{69}$$

$$w_i(t_i) = (t_i - t_0) \cdots (t_i - t_{i-1})(t_i - t_{i+1}) \cdots (t_i - t_n) \equiv \prod_{j=0, j \neq i}^{n} (t_i - t_j). \tag{70}$$

The uniform (Eq. 53),(Eq. 54) or non-uniform (Eq. 55) parametrization of a curve, together with the normalization of its domain, can be applied.

The advantages of the Lagrange interpolation technique are:

- homogeneity of control vectors,

- infinite smoothness of the resulting curve ($\mathbf{r} \in C^{\infty}$),

- computational simplicity.

However, the disadvantages of the technique are very serious. The biggest is that the norm $\parallel \mathbf{f} - \mathbf{r} \parallel$ (Eq. 59c) does not converge to 0 when n increases. In geometric terms it means increasing oscillations around the defining polygon $P_n = (\mathbf{P_0}, \mathbf{P_1}, \cdots, \mathbf{P_n})$ which can be interpreted as unfairness of $\mathbf{r}(t)$. This drawback practically eliminates the Lagrange techniques especially for bigger $n (\geq 5)$.

Lagrange's basis functions $L_i(t)$ and two examples of Lagrange interpolation are shown in Figs. 19, 20 and 21.

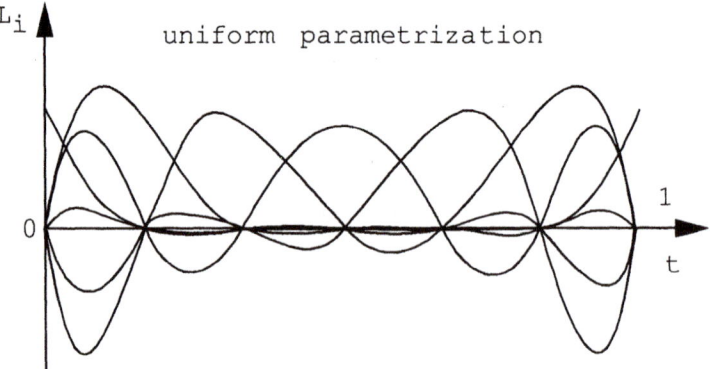

Fig. 19. Basis functions of Lagrange interpolation

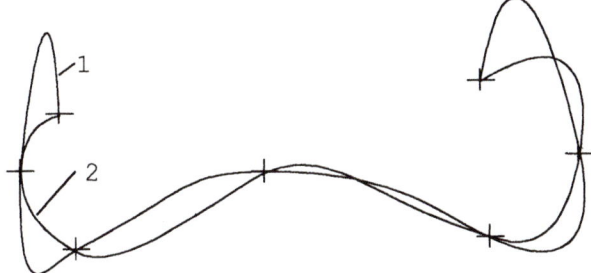

1   uniform parametrization

2   non—uniform parametrization

Fig. 20. Example of a Lagrange curve (1)

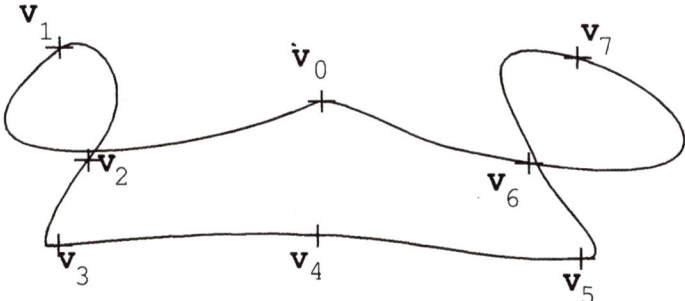

Fig. 21. Example of a Lagrange curve (2)

2.2. Hermite Interpolation

Hermite interpolation is a generalized method of Lagrange interpolation which introduces mixed interpolatory constraints (Eq. 59b) and also interpolates between higher order consecutive derivatives. The most general formulation of the Hermite technique is as follows:

"Given a knot vector $\mathbf{T} = (t_0, t_1, \cdots, t_n)$ and two corresponding sequences of positive integers: $\mathbf{k}_n = (k_0, k_1, \cdots, k_n)$ and control vectors: $\mathbf{Q}_n = \{\mathbf{Q}_{ij}\}$ $i =$

$0, 1, \cdots, n; \; j = 0, 1, \cdots, \; p_i = k_i - 1$, find a polynomial parametric curve $\mathbf{r}(t)$, (Eq. 58), of order $k = \sum_{i=0}^{n} k_i$ and degree $p = k - 1$, that solves the problem:

$$\mathbf{r}^j(t_i) = \mathbf{Q}_{ij} \quad i = 0, 1, \cdots, n; \quad j = 0, 1, \cdots, p_i. \text{"} \qquad (71)$$

There are then position vectors, tangent vectors and higher derivative vectors among $\mathbf{Q}_{ij}$. The formulation (Eq. 71) is called the full Hermite interpolation problem. Existence and uniqueness of the solution to Eq. 71 is assured by the Hermite interpolation theorem (see e.g. [64]).

It is evident that the Lagrange interpolation problem is a special case of Eq. 71, when $\mathbf{k}_n = (1, 1, 1, \cdots, 1)$, $k = n + 1, p = n$, and $\mathbf{Q}_{i0} = \mathbf{P}_i \; i = 0, 1, \cdots, n$.

The solution of the full Hermite interpolation problem (Eq. 71) is, of course, possible by solving the corresponding system of k linear equations. This is, however, a troublesome task. In CAGD practice it is quite common to take recourse to such cases only when the cardinal bases can be found directly. The two cases are discussed below:

- Full Hermite interpolation with position and tangent vectors only: $p_i = 1, i = 0, 1, 2, \cdots, n$. The data of the problem is then modified as follows:

  o a sequence of positive integers

  $$\mathbf{k}_n = (k_0, k_1, \cdots, k_n) = (2, 2, \cdots, 2),$$

  o two sequences of vectors

  $$\mathbf{P}_i, \quad i = 0, 1, \cdots, n \text{ - position vectors,}$$
  $$\mathbf{Q}_i, \quad i = 0, 1, \cdots, n \text{ - tangent vectors.}$$

The parametric polynomial to be found has order $k = 2n + 2$ and degree $p = k - 1 = 2n + 1$. The cardinal bases are correspondingly:

$$\left. \begin{array}{l} G_i(t) = (1 - 2 \cdot L_i'(t_i) \cdot (t - t_i)) \cdot L_i^2(t) \\ H_i(t) = (t - t_i) \cdot L_i^2(t) \end{array} \right\} \quad i = 0, 1, \cdots, n. \qquad (72)$$

where $L_i(t)$ is the Lagrange cardinal basis (Eq. 68).

In this case the function $\mathbf{r}(t)$ has the geometric form:

$$\mathbf{r}(t) = \sum_{i=0}^{n} \mathbf{P}_i G_i(t) + \sum_{i=0}^{n} \mathbf{Q}_i H_i(t) = [\mathbf{P}_0, \cdots, \mathbf{P}_n] \cdot \mathbf{G}(t) + [\mathbf{Q}_0, \cdots, \mathbf{Q}_n] \cdot \mathbf{H}(t). \quad (73)$$

Basis functions $G_j(t)$ and $H_j(t)$ for Hermite interpolation and two examples of Hermite interpolation (both for this case) are shown in Figs. 22, 23 and 24 respectively.

An advantage of this interpolation technique is that not only position vectors but also tangent vectors are under control. Disadvantages, however are still predominant:

o The control vectors become non-homogeneous,

o The resulting curve $\mathbf{r}(t)$ is still an ordinary polynomial (infinite times differentiable) and for $n+1$ position vectors it has the degree $p = 2n+1$ (much higher than Lagrange's).

o The curve $\mathbf{r}(t)$ tends to oscillate around the defining polygon P_n just as in the case of Lagrange technique.

To avoid these drawbacks the idea had arisen to abandon infinite differentiability and to replace full Hermite polynomial by piecewise Hermite polynomials. It has led to the second case.

- Piecewise Hermite interpolation with position and tangent vectors only: $p_i = 1$, $i = 0, 1, 2, \cdots, n$, and an additional assumption concerning the differentiability of the curve $\mathbf{r}(t)$:

o $\mathbf{r}(t) \in C^{\infty}$ between knots (locally).

o $\mathbf{r}(t) \in C^1$ at knots, so $\mathbf{r}(t) \in C^1$ (globally).

This case has been developed and applied in CAGD by Ferguson and is known as Ferguson interpolation.

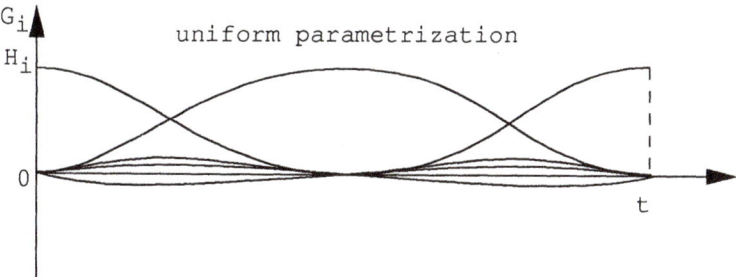

Fig. 22. Basis functions of Hermite interpolation

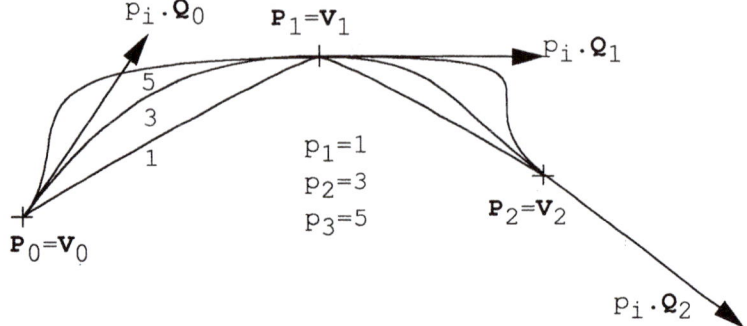

Fig. 23. Example of a Hermite curve (1)

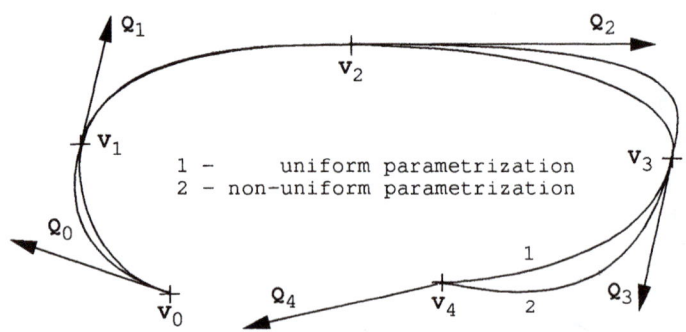

Fig. 24. Example of a Hermite curve (2)

2.3. Ferguson Interpolation

Ferguson interpolation is, on one hand, a special case of Hermite polynomial interpolation and, on the other hand, a generalization of the Hermite approach towards a $3D$ parametric curve definition.

For simplicity, let us consider first a scalar ($1D$) problem for one segment of a curve $\mathbf{r}(t)$ and a vector of knots: $\mathbf{T} = (t_0, t_1)$. The function $r(t)$ can be regarded as a component of the curve $\mathbf{r}(t)$. From Eq. 72 the cardinal basis functions are

$$\left.\begin{array}{l} G_i(t) = (1 - 2 \cdot L_i'(t_i) \cdot (t - t_i)) \cdot L_i^2(t) \\ H_i(t) = (t - t_i) \cdot L_i^2(t), \end{array}\right\} \quad i = 0, 1, \cdots, n. \tag{74}$$

If the 'control parameters' are

$$p_i = r(t_i) \quad \text{and} \quad q_i = r^1(t_i), \quad i = 0, 1 \tag{75}$$

then the segment $r(t)$ has the following geometric form:

$$r(t) = p_0 \cdot G_0(t) + p_1 \cdot G_1(t) + q_0 \cdot H_0(t) + q_1 \cdot H_1(t). \tag{76}$$

In order to find the cardinal bases $G_i(t)$ and $H_i(t)$, $i = 0, 1$, let us observe that Eq. 74 is based on the Lagrange basis functions (Eq. 68). Therefore the relations hold:

$$L_0(t) = \frac{t - t_1}{t_0 - t_1}; \quad L_0'(0) = \frac{1}{t_0 - t_1}; \quad L_0^2(t) = \frac{(t - t_1)^2}{(t_0 - t_1)^2}$$

$$L_1(t) = \frac{t - t_0}{t_1 - t_0}; \quad L_1'(1) = \frac{1}{t_1 - t_0}; \quad L_1^2(t) = \frac{(t - t_0)^2}{(t_1 - t_0)^2}. \tag{77}$$

After substituting Eq. 77 into Eq. 74 one obtains the cardinal bases for one segment of the curve $r(t)$:

$$G_0(t) = \frac{(t - t_1)^2 \cdot [2(t - t_0) + (t_1 - t_0)]}{(t_1 - t_0)^3} \tag{78}$$

$$G_1(t) = \frac{(t - t_0)^2 \cdot [2(t_1 - t) + (t_1 - t_0)]}{(t_1 - t_0)^3}$$

$$H_0(t) = \frac{(t - t_0) \cdot (t - t_1)^2}{(t_1 - t_0)^2} \tag{79}$$

$$H_1(t) = \frac{(t - t_1) \cdot (t - t_0)^2}{(t_1 - t_0)^2}.$$

For a special case of parametrization: $\mathbf{T} = (t_0, t_1) = (0, 1)$ (a unit segment of a Ferguson cubic function) one gets:

$$G_0(t) = (t - 1)^2 \cdot (2t + 1) = 1 + 0t - 3t^2 + 2t^3$$
$$G_1(t) = (3 - 2t) \cdot t^2 = 0 + 0t + 3t^2 - 2t^3$$
$$H_0(t) = t \cdot (t - 1)^2 = 0 + 1t - 2t^2 + 1t^3$$
$$H_1(t) = (t - 1) \cdot t^2 = 0 + 0t - 1t^2 + 1t^3.$$

The above can be expressed in matrix form as follows:

$$\mathbf{F}(t) \equiv \begin{bmatrix} F_1(t) \\ F_2(t) \\ F_3(t) \\ F_4(t) \end{bmatrix} = \begin{bmatrix} G_0(t) \\ G_1(t) \\ H_0(t) \\ H_1(t) \end{bmatrix} = \begin{bmatrix} 1 & 0 & -3 & 2 \\ 0 & 0 & 3 & -2 \\ 0 & 1 & -2 & 1 \\ 0 & 0 & -1 & 1 \end{bmatrix} \cdot \begin{bmatrix} 1 \\ t \\ t^2 \\ t^3 \end{bmatrix}. \tag{80}$$

This result can be used to describe one segment of Ferguson $3D$ parametric cubic curve $\mathbf{r}(t)$, $t \in [0,1]$, where the control vectors are: $\mathbf{P}_0 = \mathbf{r}(0), \mathbf{P}_1 = \mathbf{r}(1), \mathbf{Q}_0 = \mathbf{r}^1(0), \mathbf{Q}_1 = \mathbf{r}^1(1)$. For this, one obtains:

$$\mathbf{r}(t) = \begin{bmatrix} \mathbf{P}_0 \\ \mathbf{P}_1 \\ \mathbf{Q}_0 \\ \mathbf{Q}_1 \end{bmatrix}^T \cdot \begin{bmatrix} 1 & 0 & -3 & 2 \\ 0 & 0 & 3 & -2 \\ 0 & 1 & -2 & 1 \\ 0 & 0 & -1 & 1 \end{bmatrix} \cdot \begin{bmatrix} 1 \\ t \\ t^2 \\ t^3 \end{bmatrix}. \tag{81}$$

The cardinal basis (Eq. 80) for a one-segment curve can now be generalized (see Fig. 25) to define the cardinal basis for an n-segment curve with the knot vector $\mathbf{T} = (t_{-1}, t_0, t_1, \cdots, t_{n+1})$:

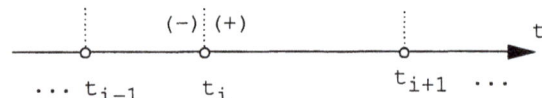

Fig. 25. The knot vector

$$G_i(t) = \begin{cases} \frac{(t_{i-1}-t)^2 \cdot [2(t_i-t)+(t_i-t_{i-1})]}{(t_i-t_{i-1})^3} & t \in [t_{i-1}, t_i] \\[2mm] \frac{(t_{i+1}-t)^2 \cdot [2(t-t_i)+(t_{i+1}-t_i)]}{(t_{i+1}-t_i)^3} & t \in [t_i, t_{i+1}] \\[2mm] 0 & \text{elsewhere,} \end{cases} \tag{82}$$

$$H_i(t) = \begin{cases} \frac{(t-t_{i-1})^2 \cdot (t-t_i)}{(t_i-t_{i-1})^2} & t \in [t_{i-1}, t_i] \\[2mm] \frac{(t-t_i) \cdot (t-t_{i+1})^2}{(t_{i+1}-t_i)^2} & t \in [t_i, t_{i+1}] \\[2mm] 0 & \text{elsewhere} \end{cases} \tag{83}$$

for $i = 0, 1, \cdots, n$.

Finally, an n-segment Ferguson curve, defined by position vectors $\mathbf{P}_i$ and tangent vectors $\mathbf{Q}_i$, $i = 0, 1, 2, \cdots, n$, takes the following geometric form:

$$\mathbf{r}(t) = \sum_{i=0}^{n} \mathbf{P}_i G_i(t) + \sum_{i=0}^{n} \mathbf{Q}_i H_i(t) = [\mathbf{P}_0, \cdots, \mathbf{P}_n] \cdot \mathbf{G}(t) + [\mathbf{Q}_0, \cdots, \mathbf{Q}_n] \cdot \mathbf{H}(t) \tag{84}$$

where $G_i(t)$ and $H_i(t)$ are cardinal bases defined by Eq. 82 and Eq. 83 respectively. Fig. 26 shows an example of the cardinal basis and Figs. 27 and 28 present examples of interpolation by the Ferguson technique.

The main advantage of the Ferguson interpolation technique is that it displays far better approximation property (the norm $\| \mathbf{f} - \mathbf{r} \|$ is less) than the Lagrange and

Hermite techniques. A Ferguson curve $r(t)$ is always the piecewise cubic polynomial curve.

The disadvantages of the Ferguson interpolation techniques are:

- The resulting piecewise cubic curve $r(t)$ has too low a degree of differentiability ($r \in C^1$).

- The control vectors are non-homogeneous vectors (position and tangent vectors together).

All of these drawbacks have been removed in the spline interpolation technique.

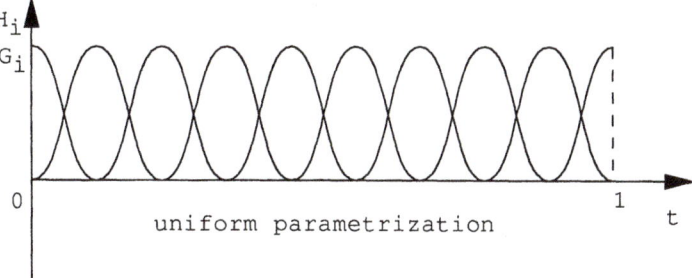

Fig. 26. Basis functions of Freguson interpolation

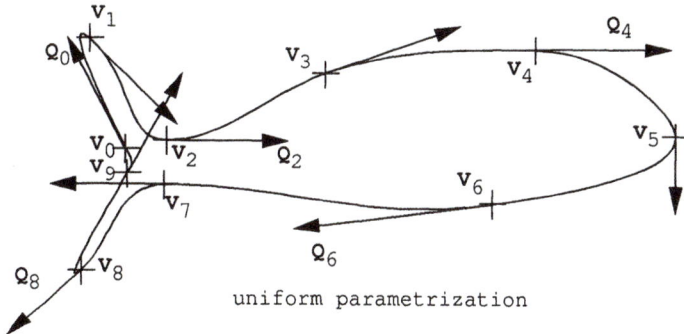

Fig. 27. Example of a Ferguson curve (1)

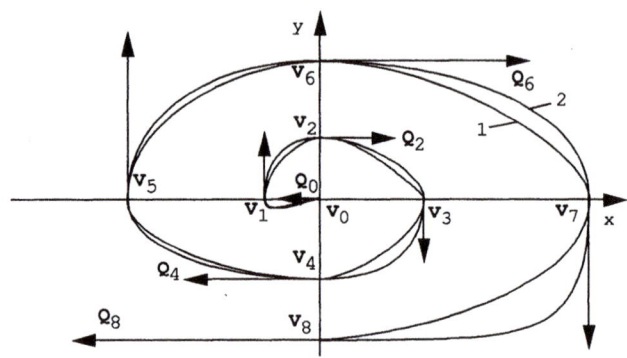

1 -   uniform parametrization
2 - non-uniform parametrization

Fig. 28. Example of a Ferguson curve (2)

2.4. Spline Interpolation

Spline interpolation techniques can be presented in a number of ways. Among others, they can directly be derived from the Ferguson interpolation results (Eqs. 82,83 and 84). A method is demonstrated below which is an example concerning the cubic spline only for the case of an open $3D$ parametric curve interpolation. The method is based on a traditional approach and does not make use of the cubic spline cardinal basis approach (introduced by Schoenberg in 1966 [64]). The idea behind this method is to eliminate the tangent vectors from the collection of control vectors through built-in C^2 conditions of smoothness of a cubic Ferguson curve $r(t)$.

For simplicity a scalar case is considered first, which will be generalized to a vector case afterwards.

- Scalar case:

As a starting point the Ferguson solution for a single-valued function $r(t)$ is taken into account:

$$r(t) = \mathbf{G}^T(t) \cdot \mathbf{p} + \mathbf{H}^T(t) \cdot \mathbf{q}; \quad t \in [t_0, t_n] \tag{85}$$

where

$$
\begin{aligned}
&\mathbf{T} = (t_{-1}, t_0, t_1, \cdots, t_n, t_{n+1}) &&\text{- knot vector} \\
&\mathbf{p} = (p_0, p_1, \cdots, p_n)^T; \;\; p_i \equiv r(t_i) &&\text{- position vector} \\
&\mathbf{q} = (q_0, q_1, \cdots, q_n)^T; \;\; q_i \equiv r^1(t_i) &&\text{- tangent vector} \\
&\mathbf{G}(t) = (G_0(t), G_1(t), \cdots, G_n(t))^T &&\text{- cardinal basis vector} \\
&\mathbf{H}(t) = (H_0(t), H_1(t), \cdots, H_n(t))^T &&\text{- cardinal basis vector.}
\end{aligned}
\tag{86}
$$

Let us introduce left- and right-side second derivatives of $r(t)$ at the knot t_i:

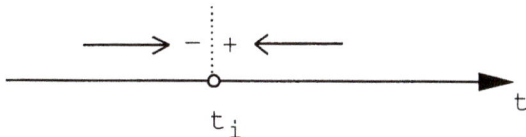

Fig. 29. Notation at the knot t_i

$$r^2_-(t_i) \equiv \frac{d^2}{dt^2}(r)\bigg|_{t=t_{i-}} \quad ; \quad r^2_+(t_i) \equiv \frac{d^2}{dt^2}(r)\bigg|_{t=t_{i+}} \qquad (87)$$

In order to find an open, cubic spline curve with an assumed C^2 smoothness s_0 and s_n at the end knots t_0 and t_n respectively and the same C^2 smoothness at the interior knots: $t_1, t_2, \cdots, t_{n-1}$, the following system of equations must be solved:

$$r^2_+(t_0) = s_0, \qquad i = 0$$

$$r^2_-(t_i) = r^2_+(t_i), \quad i = 1, 2, \cdots, n-1 \qquad (88)$$

$$r^2_-(t_n) = s_n, \qquad i = n.$$

This is the system of $N = n + 1$ linear equations with N unknowns, $\mathbf{q} = (q_0, q_1, \cdots, q_n)^T$, dim $\mathbf{q} = N$. The existence of the solution to Eq. 88 can be proven (see e.g. [64]).

Due to linearity of the system, one can expect to get a solution to Eq. 88 in the following matrix form:

$$\mathbf{q} = \mathbf{M} \cdot (\mathbf{p}; \bar{\mathbf{p}})^T = \mathbf{M_1} \cdot \mathbf{p} + \mathbf{M_2} \cdot \bar{\mathbf{p}}, \text{ where } \bar{\mathbf{p}} = (s_0, s_n)^T. \qquad (89)$$

This would give the required elimination of mixed interpolatory constraints everywhere except at the end points. The derivatives s_0, s_n at the end points can be regarded as additional 'degrees of freedom' of the spline function $r(t)$.

The linear form of the solution (Eq. 89) will be proved along with the interpretation of the matrices $\mathbf{M_1}$ and $\mathbf{M_2}$.

Thanks to the linear form of $r(t)$ (Eq. 85) the corresponding 2nd, left- and right-side derivatives of $r(t)$ at the knot t_i are:

$$r^2_{\mp}(t_i) = (\mathbf{G}^T)^2_{\mp}(t_i) \cdot \mathbf{p} + (\mathbf{H}^T)^2_{\mp}(t_i) \cdot \mathbf{q}. \qquad (90)$$

After substituting Eq. 90 into Eq. 88 and changing the initial order of Eq. 88, one obtains the system of equations:

$$\left((\mathbf{H}^T)_-^2(t_i) - (\mathbf{H}^T)_+^2(t_i)\right) \cdot \mathbf{q} +$$

$$\left((\mathbf{G}^T)_-^2(t_i) - (\mathbf{G}^T)_+^2(t_i)\right) \cdot \mathbf{p} \qquad = 0, \qquad i = 1, 2, \cdots, n-1$$

$$(\mathbf{H}^T)_+^2(t_0) \cdot \mathbf{q} + (\mathbf{G}^T)_+^2(t_0) \cdot \mathbf{p} - s_0 \; = 0, \qquad i = 0$$

$$(\mathbf{H}^T)_-^2(t_n) \cdot \mathbf{q} + (\mathbf{G}^T)_-^2(t_n) \cdot \mathbf{p} - s_n \; = 0, \qquad i = n.$$

(91)

For convenience, the following notation is introduced:

$$\mathbf{h}(t_i) \equiv \begin{cases} (\mathbf{H})_+^2(t_0), & i = 0 \\ (\mathbf{H})_-^2(t_i) - (\mathbf{H})_+^2(t_i), & i = 1, 2, \cdots, n-1 \\ (\mathbf{H})_-^2(t_n), & i = n \end{cases}$$

(92)

and correspondingly

$$\mathbf{g}(t_i) \equiv \begin{cases} (\mathbf{G})_+^2(t_0), & i = 0 \\ (\mathbf{G})_-^2(t_i) - (\mathbf{G})_+^2(t_i), & i = 1, 2, \cdots, n-1 \\ (\mathbf{G})_-^2(t_n), & i = n. \end{cases}$$

(93)

Thus, the system of equations (Eq. 91) takes the form:

$$\begin{aligned} \mathbf{h}^T(t_i) \cdot \mathbf{q} + \mathbf{g}^T(t_i) \cdot \mathbf{p} &= 0, & i = 1, 2, \cdots, n-1 \\ \mathbf{h}^T(t_0) \cdot \mathbf{q} + \mathbf{g}^T(t_0) \cdot \mathbf{p} - s_0 &= 0, & i = 0 \\ \mathbf{h}^T(t_n) \cdot \mathbf{q} + \mathbf{g}^T(t_n) \cdot \mathbf{p} - s_n &= 0, & i = n \end{aligned}$$

(94)

$$\begin{aligned} \mathbf{h}^T(t_i) \cdot \mathbf{q} &= -\mathbf{g}^T(t_i) \cdot \mathbf{p} &= 0, & i = 1, 2, \cdots, n-1 \\ \mathbf{h}^T(t_0) \cdot \mathbf{q} &= -\mathbf{g}^T(t_0) \cdot \mathbf{p} + s_0 &= 0, & i = 0 \\ \mathbf{h}^T(t_n) \cdot \mathbf{q} &= -\mathbf{g}^T(t_n) \cdot \mathbf{p} + s_n &= 0, & i = n. \end{aligned}$$

(95)

The latter one can be expressed in a compact matrix notation:

$$\mathbf{H} \cdot \mathbf{q} = \mathbf{G} \cdot (\mathbf{p}, \overline{\mathbf{p}})^T$$

(96)

where $\overline{\mathbf{p}} = (s_0, s_n)^T$ - a pair of 2nd derivatives at the end points;

$$\mathbf{H} = \left[\; h_k(t_i) \;\mid\; h_k(t_0), h_k(t_n) \;\right]^T$$

(97)

$$\mathbf{G} = -\begin{bmatrix} g_k(t_i) & \mid & g_k(t_0), g_k(t_n) \\ --- & \mid & ----- \\ 0 & \mid & -I \end{bmatrix}^T$$

(98)

for $k = 0, 1, \cdots, n$, $i = 1, 2, \cdots, n-1$ and det $\mathbf{H} \neq 0$, $I \equiv \text{diag}(1)$.

Finally, the solution of the matrix equation (Eq. 96), with respect to q is

$$\mathbf{q} = \mathbf{H}^{-1} \cdot \mathbf{G} \cdot (\mathbf{p}, \overline{\mathbf{p}})^T = \mathbf{M} \cdot (\mathbf{p}, \overline{\mathbf{p}})^T = \mathbf{M}_1 \cdot \mathbf{p} + \mathbf{M}_2 \cdot \overline{\mathbf{p}} \qquad (99)$$

where $\mathbf{M}_1$ and $\mathbf{M}_2$ are two corresponding sub-matrices of $\mathbf{M}$. It can be shown that the matrix $\mathbf{H}$ is a three-diagonal matrix. It follows from Gershgorin's theorem [64] that such a matrix is nonsingular, and hence a unique solution to Eq. 99 will always exist.

In such a way the solution to the system (Eq. 88), in the form of Eq. 89, for the scalar case, has been proven.

Substituting Eq. 99 into Eq. 85 yields

$$
\begin{aligned}
r(t) = \ & \mathbf{G}^T(t) \cdot \mathbf{p} + \mathbf{H}^T(t) \cdot \mathbf{q} \ = \mathbf{G}^T(t) \cdot \mathbf{p} + \mathbf{H}^T(t) \cdot (\mathbf{M}_1 \cdot \mathbf{p} + \mathbf{M}_2 \cdot \overline{\mathbf{p}}) \\
& = \mathbf{G}^T(t) \cdot \mathbf{p} + \mathbf{H}^T(t) \cdot \mathbf{M}_1 \cdot \mathbf{p} + \mathbf{H}^T(t) \cdot \mathbf{M}_2 \cdot \overline{\mathbf{p}} \\
& = \underbrace{\left(\mathbf{G}^T(t) + \mathbf{H}^T(t) \cdot \mathbf{M}_1 \right)}_{\overline{\mathbf{G}}^T(t)} \cdot \mathbf{p} + \underbrace{\left(\mathbf{H}^T(t) \cdot \mathbf{M}_2 \right)}_{\overline{\mathbf{H}}^T(t)} \cdot \overline{\mathbf{p}}.
\end{aligned}
$$

$$(100)$$

The two 'new' cardinal bases have been defined above. As a result

$$r(t) = \overline{\mathbf{G}}^T(t) \cdot \mathbf{p} + \overline{\mathbf{H}}^T(t) \cdot \overline{\mathbf{p}}. \qquad (101)$$

For $\overline{\mathbf{p}} = (0, 0)^T$, a spline $r(t)$ is called a natural spline. This is the best mathematical model of a physical spline. The result (Eq. 101) can now be generalized to get:

- Vector case:

A vector case for the parametric cubic spline interpolation technique is instantly obtained from Eq. 101:

$$
\begin{aligned}
\mathbf{r}(t) \ & = \ [\mathbf{P}_0, \mathbf{P}_1, \cdots, \mathbf{P}_n] \cdot \overline{\mathbf{G}}(t) + [\overline{\mathbf{P}}_0, \overline{\mathbf{P}}_n] \cdot \overline{\mathbf{H}}(t) \\
& = \ \mathbf{P} \cdot \overline{\mathbf{G}}(t) + \overline{\mathbf{P}} \cdot \overline{\mathbf{H}}(t).
\end{aligned}
\qquad (102)
$$

This form contains homogeneous position vectors $\mathbf{P}_0, \mathbf{P}_1, \cdots, \mathbf{P}_n$ at all the internal knots and, additionally, two 2nd derivatives vectors $\overline{\mathbf{P}}_0$ and $\overline{\mathbf{P}}_n$ at the end knots. The curve $\mathbf{r}(t)$ as a cubic spline is C^2 - continuous at the internal knots and C^∞ - continuous between knots.

Three examples of the cubic spline interpolation technique, based on the solution (Eq. 102) are shown in Figs. 30, 31 and 32.

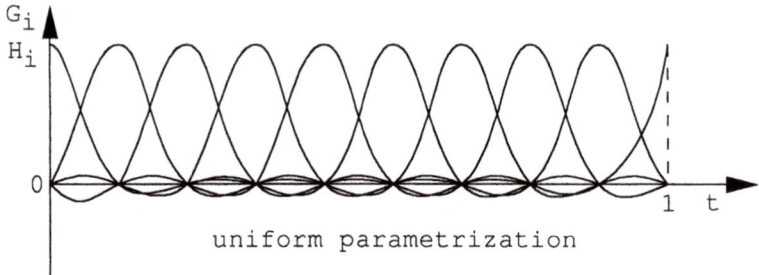

Fig. 30. Basis functions of cubic-spline interpolation

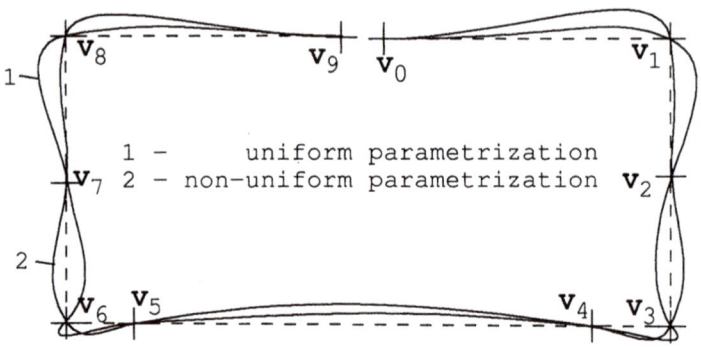

Fig. 31. Example of a cubic-spline curve (1)

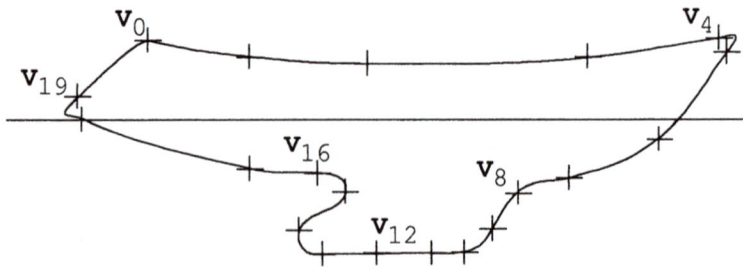

Fig. 32. Example of a cubic-spline curve (2)

3. Control Polygon Techniques

In general, the idea of control polygon techniques is based on the approximation problem (see Section 1.5), and so these techniques are sometimes called approximation techniques. In CAGD, approximation techniques have been developed to form a very modern and powerful tool of geometric modelling [49].

In all of these approximation techniques a curve (function) to be approximated is given in a discrete manner, by a sequence of control vectors $\mathbf{V}_i$, $i = 0, 1, 2, \cdots, n$, which are exclusively position vectors. As in the interpolation techniques, all the neighbouring position vectors connected to each other by straight lines form a defining polygon $P_n = (\mathbf{V}_0, \mathbf{V}_1, \cdots, \mathbf{V}_n)$ with vertices $\mathbf{V}_i$ and edges $\overline{\mathbf{V}_i\mathbf{V}_{i+1}}$, $i = 0, 1, \cdots, n-1$. In the case of approximation, the defining polygon P_n can be regarded as a piecewise linear function $\mathbf{f}(t)$ to be approximated. However, in contrast to interpolation techniques, in approximation techniques the curve $\mathbf{r}(t)$, to be found, is not forced to pass exactly through the vertices of P_n. If $\mathbf{P}_i$ are points on the curve: $\mathbf{P}_i = \mathbf{r}(t_i)$, then, in general, $\mathbf{V}_i \neq \mathbf{P}_i$. The idea is illustrated in Fig. 33.

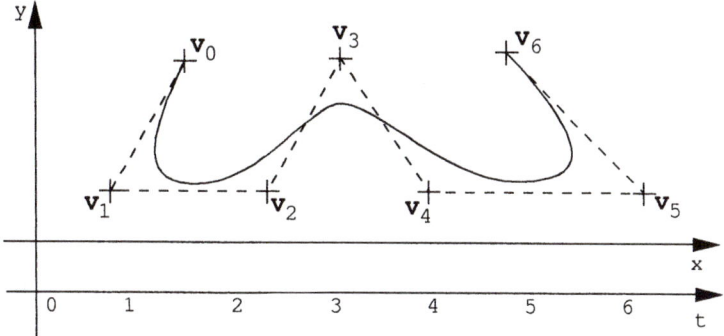

Fig. 33. Approximation technique - control vectors

In approximation techniques the geometric form of a curve definition (Eq. 65) is commonly used and are of the general form:

$$\mathbf{r}(t) = \sum_{i=0}^{n} \mathbf{V}_i \cdot F_i(t) \tag{103}$$

where $F_i(t)$ are blending (basis) functions. The blending functions distinguish different approximation techniques and entirely determine their properties as well as their usefulness in CAGD.

Three specific control polygon (approximation) techniques will be discussed below; Bézier, B-Spline, and its extension - Non Uniform Rational B-Splines (NURBS), approximation techniques. In all techniques, the blending functions F_i are, in general, based on polynomial functions.

3.1. Bézier Approximation

Curves created by Bézier approximation techniques are called Bézier curves and vertices V_i of its defining polygon - Bézier points. The Bézier technique results in the approximation of a function by Bernstein polynomials.

3.2. Bernstein Polynomials

Bernstein polynomials result from of Bernstein approximations. An arbitrary continuous function $f(t)$, $t \in [0,1]$ can be approximated by an nth degree Bernstein polynomial $B_n(t)$ in the form:

$$B_n(t) = \sum_{i=0}^{n} B_{i,n}(t) \cdot f(\frac{i}{n}), \quad t \in [0,1] \tag{104}$$

where $B_{i,n}(t)$ are Bernstein basis functions defined as follows:

$$B_{i,n}(t) = \left(\begin{array}{c} n \\ i \end{array} \right) \cdot t^i \cdot (1-t)^{n-i} \quad i \in [0,n], t \in [0,1]$$

$$B_{i,n}(t) \equiv 0 \qquad\qquad\qquad i < 0 \text{ or } i > n. \tag{105}$$

An example of the Bernstein basis function is shown in Fig. 34.

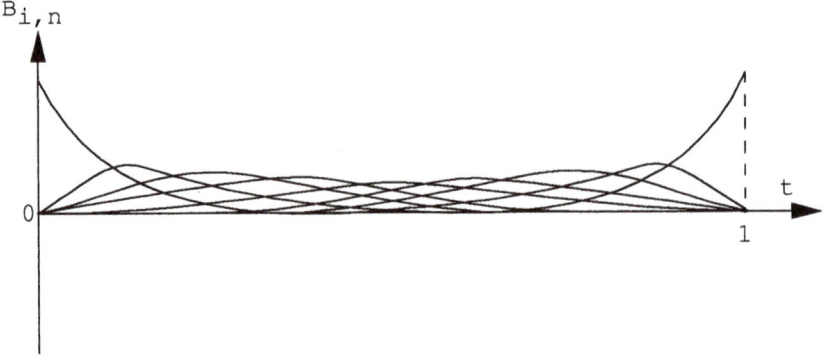

Fig. 34. The Bernstein basis functions

The Bernstein basis functions $B_{i,n}(t)$ have some important properties. They influence the Bernstein polynomials $B_n(t)$ and are the key to the behaviour of Bézier curves. The basic properties of the Bernstein basis functions and the Bernstein polynomials are surveyed in Tables 1 and 2.

It follows from the convergence (Eq. 106) and boundedness (Eq. 107) properties that if $f(t)$ is monotonic (convex, concave) then $B_n(t)$ is also monotonic (convex,

Table 1. Properties of the Bernstein basis functions

Property	Analytical form
Positivity	$B_{i,n}(t) \geq 0, \quad i = 0, 1, \cdots, n \quad t \in [0, 1]$ $B_{i,n}(t) < 1, \quad i = 1, 2, \cdots, n - 1$ $B_{i,n}(t) \leq 1, \quad i = 0, 1$
Cauchy's relation (partition of unity)	$\sum_{i=0}^{n} B_{i,n}(t) = 1$
Recursion	$B_{i,n}(t) = (1 - t) \cdot B_{i,n-1}(t) + t \cdot B_{i-1,n-1}(t)$ $(B_{i,n})^{1}(t) = n \cdot (B_{i-1,n-1}(t) - B_{i,n-1}(t))$
Symmetry	$B_{n-i,n}(t) = B_{i,n}(t) \cdot (1 - t)$

Table 2. Properties of the Bernstein polynomials

Property	Analytical form
Convergence	$f(t) \in [a, b] \quad \Rightarrow \quad \lim B_n(t) = f(t)$ <div align="right">(106)</div>
Boundedness	$f^k(t) \in [a_k, b_k] \quad \Rightarrow$ $B_n(t) \in [a_k, b_k] \quad k = 0$ $\frac{n^k \cdot (B_n)^k(t)}{n(n-1)\cdots(n-k+1)} \in [a_k, b_k] \quad k \in [1, n]$ <div align="right">(107)</div>
Linearity	$f(t) = at + b \Rightarrow B_n(t) = at + b$ <div align="right">(108)</div>
Variation Diminishing Property	$S(B_n(t) - at - b) \leq S(f(t) - at - b)$ $S(f)$ - a number of sign changes of $f(t)$ in $[0, 1]$. <div align="right">(109)</div>

concave). That is, the Bernstein approximation expresses the overall form of $f(t)$ very well. The linearity property assures that the Bernstein approximation of a linear function is the linear function itself.

The geometrical significance of the inequality (Eq. 109) is that the number of intersections of a function $y = at + b$ with $y = B_n(t)$ does not exceed the number of intersections of $y = at + b$ with $y = f(t)$. This property is called the variation diminishing property of a Bernstein polynomial. That is to say, if a Bernstein approximation is applied to $f(t)$ a smoothing effect is obtained. After an infinite number of approximation repetitions, the function $B_n(t)$ reduces to the straight line joining $f(0)$ and $f(1)$.

3.3. Definition of a Bézier Curve

A Bézier curve is a $3D$ version of a Bernstein approximation. An approximated function is meant to be a vector-valued function $\mathbf{f}(t)$ forming a characteristic polygon P_n. The Bernstein polynomial serves as a blending (basis) functions $F_i(t)$ in Eq. 103.

A Bézier curve of pth degree , where $p = n$, has the following general form

$$\mathbf{r}(t) = \sum_{i=0}^{n} \mathbf{V}_i \cdot F_i(t) = \sum_{i=0}^{n} \mathbf{V}_i \cdot B_{i,n}(t) = [\mathbf{V}_0, \mathbf{V}_1, \cdots, \mathbf{V}_n] \cdot \mathbf{B}_n = \mathbf{V} \cdot \mathbf{B}_n$$

$$t \in [0,1]. \quad (110)$$

Eq. 110 describes one Bézier curve segment. In particular, for the most applicable case - a Bézier cubic segment, one gets the following basis:

$$
\begin{aligned}
F_0(t) &\equiv B_{0,3}(t) = (1-t)^3 = 1 - 3t + 3t^2 - t^3 \\
F_1(t) &\equiv B_{1,3}(t) = 3t \cdot (1-t)^2 = 0 + 3t - 6t^2 + 3t^3 \\
F_2(t) &\equiv B_{2,3}(t) = 3t^2 \cdot (1-t) = 0 + 0 + 3t^2 - 3t^3 \\
F_3(t) &\equiv B_{3,3}(t) = t^3 = 0 + 0 + 0 + t^3.
\end{aligned}
\quad (111)
$$

The corresponding Bézier cubic segment can compactly be written in the following matrix form and in the ordinary power basis:

$$
\mathbf{r}(t) =
\begin{bmatrix} \mathbf{V}_0 \\ \mathbf{V}_1 \\ \mathbf{V}_2 \\ \mathbf{V}_3 \end{bmatrix}^T
\cdot
\begin{bmatrix} 1 & -3 & 3 & -1 \\ 0 & 3 & -6 & 3 \\ 0 & 0 & 3 & -3 \\ 0 & 0 & 0 & 1 \end{bmatrix}
\cdot
\begin{bmatrix} 1 \\ t \\ t^2 \\ t^3 \end{bmatrix} . \quad (112)
$$

One can compare Eq. 112 with that of a Ferguson cubic segment (Eq. 81) to see the difference.

Eq. 112 can be generalized to get an nth degree Bézier curve in the power basis (Eq. 57) instead of a Bernstein basis $B_{i,n}(t)$:

$$r(t) = [\mathbf{V}_0, \mathbf{V}_1, \cdots, \mathbf{V}_n] \cdot \mathbf{B}^T \cdot \mathbf{E}(t) \tag{113}$$

where the matrix $\mathbf{B} = [b_{ij}]$ $i, j = 0, 1, \cdots, n$ according to [23], is:

$$b_{ij} = \begin{cases} (-1)^{n-i-j} \begin{pmatrix} n \\ n-i \end{pmatrix} \begin{pmatrix} n-i \\ j \end{pmatrix}, & i+j \in [0, n] \\ 0 & \text{otherwise.} \end{cases} \tag{114}$$

3.4. Properties of Bézier Curves

The properties of Bernstein polynomials and Bernstein approximations listed above result in some geometric properties of a Bézier curve. From a practical point of view, which means simplicity and convenience when applying Bézier curves in CAGD, some of their properties are very positive (desirable), some of them, however, are rather negative (undesirable).

The positive properties (global and local) of Bézier curves are:

o The Geometry Invariant Property (global):

The shape of a Bézier curve is determined only by its vertices and is not related to the coordinate system. Moreover, Cauchy's relation (Table 1) assures the shape of a curve to be invariant under coordinate transformation.

o The Variation Diminishing Property (global):

No plane has more intersections with the Bézier curve than with its characteristic polygon. This means that the Bézier curve never oscillates beyond the polygon and preserves its principal feature of the shape.

o The Convex Hull Property (global):

The positivity of basis and linearity of approximation guarantees that the Bézier curve lies completely within the convex hull of vertices $\mathbf{V}_i$. In particular, if all $\mathbf{V}_i$ are collinear, the Bézier curve is the straight line: $\mathbf{V}_0 - \mathbf{V}_n$. If all $\mathbf{V}_i$ coincide, the Bézier curve degenerates to a single point.

o The Symmetry Property (global):

Setting a renumbering of the vertices: $\overline{\mathbf{V}}_{n-i} = \mathbf{V}_i$ ensures that the following identity holds:

$$\sum_{i=0}^{n} \overline{\mathbf{V}}_i \cdot F_i(t) \equiv \sum_{i=0}^{n} \mathbf{V}_i \cdot F_i(1-t). \tag{115}$$

This shows that the two Bézier curves determined alternatively by two polygons $\mathbf{V}_i$ or $\overline{\mathbf{V}}_i$ are the same in shape.

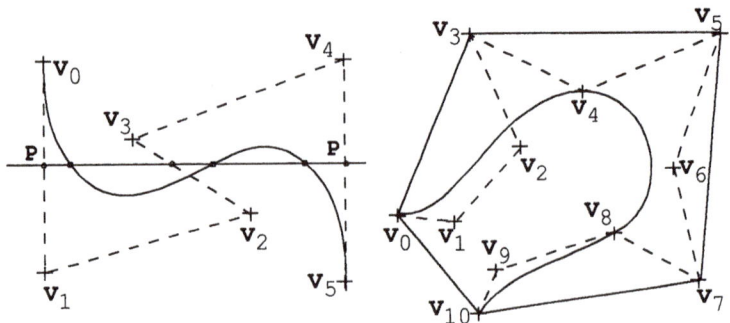

Fig. 35. The variation diminishing and convex hull property

o The End Points Property (local):

From the definition of Bernstein polynomials and a Bézier curve (Eq. 105),(Eq. 110) and from the recursion formula (Table 1) one gets:

a.

$$\mathbf{r}(0) = \mathbf{V}_0$$
$$\mathbf{r}(1) = \mathbf{V}_1 \tag{116}$$

b.

$$\mathbf{r}^k(0) = \frac{n!}{(n-k)!} \sum_{i=0}^{k} (-1)^{k-i} \cdot \begin{pmatrix} k \\ i \end{pmatrix} \cdot \mathbf{V}_i$$
$$\mathbf{r}^k(1) = \frac{n!}{(n-k)!} \sum_{i=0}^{k} (-1)^{i} \cdot \begin{pmatrix} k \\ i \end{pmatrix} \cdot \mathbf{V}_{n-i}. \tag{117}$$

In particular

c.

$$\mathbf{r}^1(0) = n \cdot (\mathbf{V}_1 - \mathbf{V}_0)$$
$$\mathbf{r}^1(1) = n \cdot (\mathbf{V}_n - \mathbf{V}_{n-1}) \tag{118}$$

d.

$$\mathbf{r}^2(0) = n \cdot (n-1) \cdot [(\mathbf{V}_2 - \mathbf{V}_1) - (\mathbf{V}_1 - \mathbf{V}_0)]$$
$$\mathbf{r}^2(1) = n \cdot (n-1) \cdot [(\mathbf{V}_n - \mathbf{V}_{n-1}) - (\mathbf{V}_{n-1} - \mathbf{V}_{n-2})] \tag{119}$$

e. The Binormal vectors

$$\mathbf{B}(0) = \mathbf{r}^1(0) \times \mathbf{r}^2(0) = n^2 \cdot (n-1) \cdot (\mathbf{V}_1 - \mathbf{V}_0) \times (\mathbf{V}_2 - \mathbf{V}_1)$$
$$\mathbf{B}(1) = \mathbf{r}^1(1) \times \mathbf{r}^2(1) = n^2 \cdot (n-1) \cdot (\mathbf{V}_{n-1} - \mathbf{V}_{n-2}) \times (\mathbf{V}_n - \mathbf{V}_{n-1}) \tag{120}$$

Properties **a.** and **c.** imply that a Bézier curve always starts on the first vertex V_0, is tangent to the first edge of the polygon: $V_0 - V_1$ and always ends on the last vertex V_n, being tangent to the last edge of the polygon: $V_n - V_{n-1}$. Moreover, **e.** implies that the corresponding osculating planes of the curve are defined by the first two or the last two edges of the polygon.

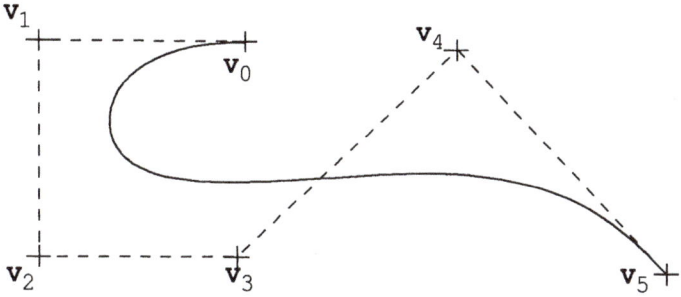

Fig. 36. The end points property

o The Multiple Vertices Property (local)

It can be shown that a multiple vertex will make the Bézier curve 'pull' in closer and closer to the space position of the multiple vertex. To do this, the degree of the Bézier curve has to be increased. This provides a useful tool to change the shape of the curve without changing its characteristic polygon.

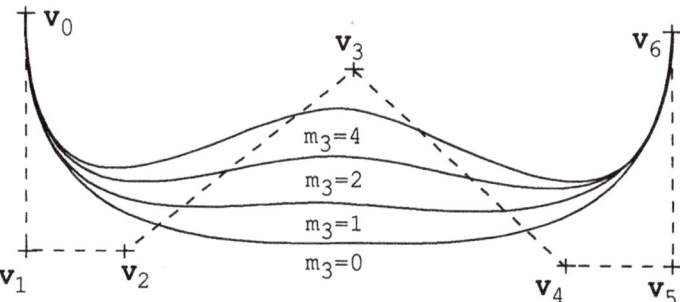

Fig. 37. The multiple vertices property

The negative properties (global and local) of Bézier curves are:

o The Order of a Curve Property (global):

The order k (or degree $p = k - 1$) of a Bézier curve is dependent on the given number of vertices (k=n+1, p=n). The advantage of higher order Bézier curve is that one can achieve a more complex curve shape and correspondingly higher order of continuity between adjacent segments of compound curve. However, the higher the degree, the more undulating the curve is likely to be and the shape relationships between the curve and the polygon becomes weaker. This is why the maximum degree of the curve, applicable in CAGD practice, should not exceed 7 [23].

o The Control Shape Property (global):

A change of one vertex changes the whole Bézier curve and re-computation of all points on the curve is required.

3.5. Algorithms of Bézier Curves

Algorithms of a Bézier curve serve for manipulation of the shape in the process of geometric modelling of the curve.

o The Degree Elevation Algorithm

The algorithm serves for increasing the number of vertices n, and the degree $p = n$ of a Bézier curve from n to $n + 1$ without changing the shape of the curve. It can be proven that, if a new set of vertices is set up, according to the formula:

$$\left.\begin{array}{l} \overline{\mathbf{V}}_i = \lambda_i \cdot \mathbf{V}_{i-1} + (1 - \lambda_i) \cdot \mathbf{V}_i \\[2mm] \lambda_i = \frac{i}{n+1} \end{array}\right\} \quad i = 0, 1, 2, \cdots, n + 1 \qquad (121)$$

then the following identity holds:

$$\overline{\mathbf{r}}(t) = \sum_{i=0}^{n} \overline{\mathbf{V}}_i \cdot B_{i,n+1}(t) \equiv \sum_{i=0}^{n} \mathbf{V}_i \cdot B_{i,n}(t) = \mathbf{r}(t). \qquad (122)$$

o The Subdivision Algorithm

The algorithm serves for partitioning of a Bézier curve segment. If we let $\mathbf{r}(t; \mathbf{V}_0, \mathbf{V}_1, \cdots, \mathbf{V}_n)$ be a given Bézier curve ($t \in [0, 1]$) and a parameter $t = t_s$ is chosen which splits up the curve $\mathbf{r}(t)$ into two new segments of the same degree but with a new parametrization:

$$\overline{\mathbf{r}}(\overline{t}; \overline{\mathbf{V}}_0, \overline{\mathbf{V}}_1, \cdots, \overline{\mathbf{V}}_n), \quad t \in [0, t_s], \quad \overline{t} = \frac{t}{t_s} \Rightarrow \overline{t} \in [0, 1] \qquad (123)$$

$$\dot{\mathbf{r}}(\dot{t}; \dot{\mathbf{V}}_0, \dot{\mathbf{V}}_1, \cdots, \dot{\mathbf{V}}_n), \quad t \in [t_s, 1], \quad \dot{t} = \frac{t - t_s}{1 - t_s} \Rightarrow \dot{t} \in [0, 1], \tag{124}$$

it can be proven, that the new segments are then expressed:

- The first segment:

$$\overline{\mathbf{r}}(t) = [\overline{\mathbf{V}}_0, \overline{\mathbf{V}}_1, \cdots, \overline{\mathbf{V}}_n] \cdot \mathbf{B}^T \cdot \mathbf{E}(t) \tag{125}$$

where the new vertices are defined as follows:

$$\overline{\mathbf{V}}_k = \sum_{j=0}^{k} B_k(t_s) \cdot \mathbf{V}_j \quad k \in [0, n]. \tag{126}$$

- The second segment:

$$\dot{\mathbf{r}}(t) = [\dot{\mathbf{V}}_0, \dot{\mathbf{V}}_1, \cdots, \dot{\mathbf{V}}_n] \cdot \mathbf{B}^T \cdot \mathbf{E}(t), \tag{127}$$

with the new vertices:

$$\dot{\mathbf{V}}_k = \sum_{j=0}^{n-k} B_{n-k}(t_s) \cdot \mathbf{V}_{k+j} \quad k \in [0, n]. \tag{128}$$

In both formulae (Eq. 125 and Eq. 127), the matrix $\mathbf{B}$ is as in Eq. 114.

o The Connection Algorithm

The algorithm serves for adding a new Bézier curve to an existing one to form a composite Bézier curve consisting of two segments. The solution is shown which is based on a cubic Bézier curve segments with C^2 parametric continuity. This degree of continuity includes:

C^0 - positional continuity - common position vertices at a joint
C^1 - gradient continuity - common tangent line at a joint
C^2 - curvature continuity - common curvature centre at a joint.

If $\mathbf{r}(t)$, $t \in [0, 1]$ is an existing curve and $\overline{\mathbf{r}}(\overline{t})$, $\overline{t} \in [0, 1]$ is a curve to be connected to $\mathbf{r}(t)$ at $t = 1 \equiv \overline{t} = 0$ then the continuity conditions result in the following system of equations:

$$\begin{aligned} \overline{\mathbf{r}}^0(0) &= \mathbf{r}^0(1) \\ \overline{\mathbf{r}}^1(0) &= c_1 \cdot \mathbf{r}^1(1), \quad c_1 \geq 0 \\ \overline{\mathbf{r}}^2(0) &= c_2 \cdot \mathbf{r}^2(1), \quad c_2 \geq 0 \end{aligned} \tag{129}$$

The solution to the system (Eq. 129) is

$$\begin{aligned}
\overline{\mathbf{V}}_0 &= \mathbf{V}_3 \\
\overline{\mathbf{V}}_1 &= c_1 \cdot (\mathbf{V}_3 - \mathbf{V}_2) + \mathbf{V}_3 \\
\overline{\mathbf{V}}_2 &= c_1 \cdot \mathbf{V}_1 - 2(c_1 + c_2) \cdot \mathbf{V}_2 + (1 + 2c_1 + c_2) \cdot \mathbf{V}_3.
\end{aligned} \tag{130}$$

This means that some constraints on $\bar{\mathbf{r}}(t)$ are imposed which change a number (from 4 to 3) and the type of its degree of freedom (no longer homogeneous position vectors). The three vertices $\mathbf{V}_2, \mathbf{V}_3 = \overline{\mathbf{V}}_0, \overline{\mathbf{V}}_1$ must be collinear, and the five $\mathbf{V}_1, \mathbf{V}_2, \mathbf{V}_3 = \overline{\mathbf{V}}_0, \overline{\mathbf{V}}_1, \overline{\mathbf{V}}_2$ must be coplanar. This also means that the osculating planes of the two curves at their joint must coincide.

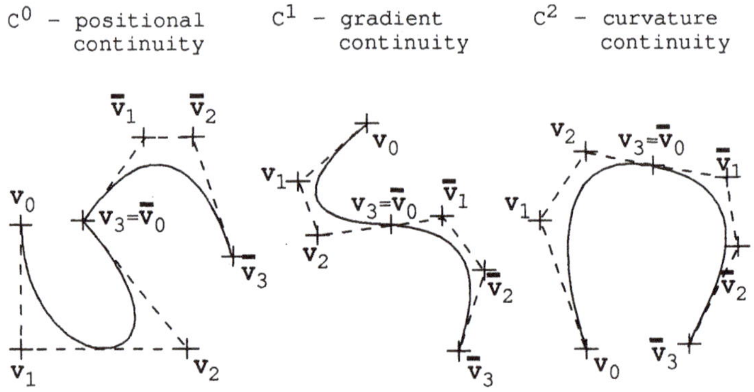

c^0 – positional continuity $\qquad$ c^1 – gradient continuity $\qquad$ c^2 – curvature continuity

Fig. 38. Composite Bézier curves

o The Positive Algorithm

Algorithms for the calculation of $r(t)$ at a given value $t = t_s$ are called the positive (or normal) algorithms. In the case of Bézier curves, a standard positive algorithm is the de Casteljau algorithm which follows directly from the recursion property (Table 1) of the Bernstein basis. In the de Casteljau algorithm a set of auxiliary vertices $\mathbf{V}_{ij}$ is recursively calculated which converges to a point $\mathbf{P}_s$ on a Bézier curve. The algorithm is as follows:

$$\mathbf{V}_{i,0}(t_s) = \mathbf{V}_i, \quad i = 0, 1, 2, \cdots, n;$$

$$\mathbf{V}_{i,j}(t_s) = (1 - t_s) \cdot \mathbf{V}_{i,j-1} + t_s \cdot \mathbf{V}_{i+1,j-1} \tag{131}$$

$$j = 1, 2, \cdots, n; \quad i = 0, 1, 2, \cdots, n - j;$$

where i denotes the order of auxiliary vertex, j the order of recursion. It can be proven [23] that

$$\begin{aligned}
\mathbf{P}_s &= \mathbf{r}(t_s) = \mathbf{V}_{0,n}(t_s) & \text{a point on a curve} \\
\mathbf{r}^1(t_s) &= n \cdot (\mathbf{V}_{1,n-1}(t_s) - \mathbf{V}_{0,n-1}(t_s)) & \text{a tangent vector at } \mathbf{P}(t_s).
\end{aligned} \qquad (132)$$

A plotting procedure (Fig. 39) illustrates the de Casteljau algorithm.

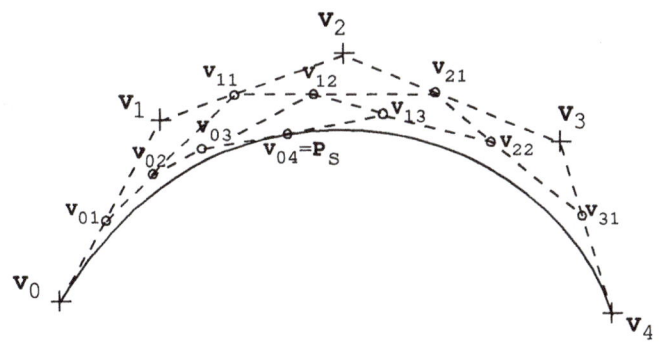

Fig. 39. Interpretation of the de Casteljau algorithm

o The Inverse Algorithm

The Bézier approach defines a curve $\mathbf{r}(t)$ that only approximates to the given (control) points $\mathbf{V}_i$. As a result, a sequence of points $\mathbf{P}_j = \mathbf{r}(t_j)$ on the curve is obtained. In CAGD, however, it is sometimes desirable to solve an inverse problem: to find a set of vertices $\mathbf{V}_i$ and then a Bézier curve while a set of points $\mathbf{P}_j$ on the curve is given. This is, in fact, an interpolation problem. The inverse algorithm serves for solving this problem. Due to a linear structure of $\mathbf{r}(t)$, the inverse algorithm leads to the solution of a system of linear equations.

According to Eq. 110, a Bézier curve has the following general form

$$\mathbf{r}(t) = \sum_{i=0}^{n} \mathbf{V}_i \cdot B_{i,n}(t) = [\mathbf{V}_0, \mathbf{V}_1, \cdots, \mathbf{V}_n] \cdot \mathbf{B}_n = \mathbf{V} \cdot \mathbf{B}_n$$

$$t \in [0,1]. \qquad (133)$$

Let $\mathbf{T} = (t_0, t_1, \cdots, t_n)$ be a vector of knots, and $\mathbf{P}_j = \mathbf{r}(t_j)$ $j = 0, 1, \cdots, n$ be a set of points to be interpolated. The interpolation problem then reduces to the following system of linear equations, with respect to $\mathbf{V}_i$:

$$\sum_{i=0}^{n} \mathbf{V}_i \cdot B_{i,n}(t_j) = \mathbf{P}_j \quad j = 0, 1, \cdots, n. \tag{134}$$

The system (Eq. 134) can be written in the following matrix form:

$$\mathbf{V} \cdot \mathbf{B} = \mathbf{P} \tag{135}$$

where

$$
\begin{aligned}
\mathbf{P} &= [\mathbf{P}_0, \mathbf{P}_1, \cdots, \mathbf{P}_n], \quad \mathbf{P}_j \equiv \mathbf{r}(t_j) \\
\mathbf{V} &= [\mathbf{V}_0, \mathbf{V}_1, \cdots, \mathbf{V}_n] \\
\mathbf{B} &= [B_{i,n}(t_j)] \quad i = 0, 1, \cdots, n; \quad j = 0, 1, \cdots, n.
\end{aligned} \tag{136}
$$

It can be proven that, since the Bernstein polynomials $B_{i,n}(t)$ are the basis functions, then the matrix $\mathbf{B}$ is nonsingular, so $\det \mathbf{B} \neq 0$. Therefore the system (Eq. 135) has a unique solution:

$$\mathbf{V} = \mathbf{P} \cdot \mathbf{B}^{-1}. \tag{137}$$

Substituting Eq. 137 into Eq. 110 one gets the interpolating Bézier curve:

$$\mathbf{r}(t) = \mathbf{V} \cdot \mathbf{B}_n(t) = \mathbf{P} \cdot \mathbf{B}^{-1} \cdot \mathbf{B}_n(t) = \mathbf{P} \cdot \mathbf{F}_n(t) \tag{138}$$

where $\mathbf{F}_n(t)$ can be regarded as a cardinal basis of Bézier interpolation (compare with Eqs. 63 and 65).

Examples of Bézier curves are shown in Figs. 40-43.

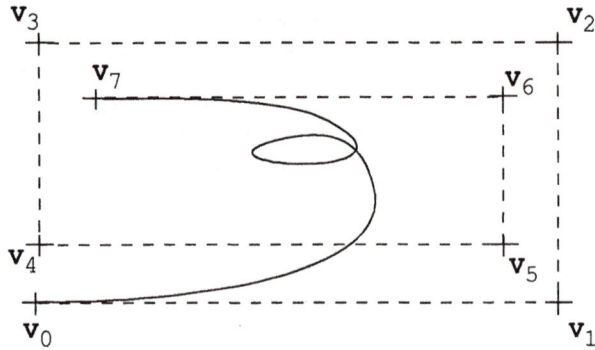

Fig. 40. Example of a Bézier curve (1)

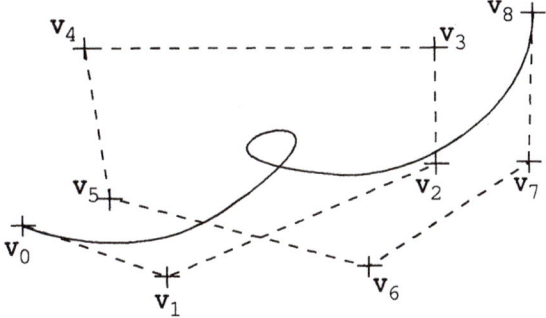

Fig. 41. Example of a Bézier curve (2)

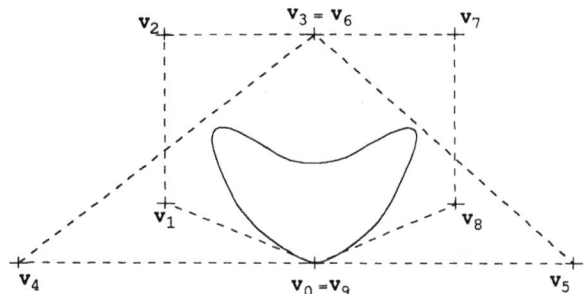

Fig. 42. Example of a Bézier curve (3)

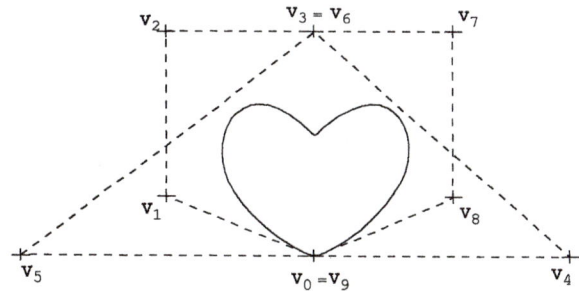

Fig. 43. Example of a Bézier curve (4)

3.6. B-Spline Approximation

B-spline curves, within the context of approximation techniques, are the result of a further intensive development of the Bézier method. The idea was to replace the Bernstein basis, used in the Bézier approximation, by a polynomial spline basis. The modern mathematical theory of spline approximation was introduced by I.J.Schoenberg in 1946. The application of this theory in CAGD has made it possible, among other things, to avoid the drawbacks of Bézier curves; thanks to their excellent geometric properties, B-spline curves have become a standard tool of computational geometry and geometric modelling nowadays [10].

All spline functions, like ordinary polynomials, form their own linear space. This means that they can be expressed by a linear combination of basis functions. Among many bases of spline functions, a basis called the B-spline basis is of the greatest importance in CAGD since it is the essence of B-spline functions.

3.7. B-Spline Basis

The importance of the B-spline basis stems from its exceptional properties. It is also interesting as the corresponding spline extension of the Bernstein basis, the mathematical underpinning of Bézier curves.

B-spline bases can be derived and expressed in different ways. The most convenient and general definition was given by de Boor, Cox and Mansfield [27]. It has the following, recursive form and expresses the normalized B-spline basis as:

$$N_{i,1}(t) = \begin{cases} 1 & \text{for } t \in [t_i, t_{i+1}) \\ 0 & \text{otherwise} \end{cases}$$

$$N_{i,j}(t) = \frac{t-t_i}{t_{i+j-1}-t_i} \cdot N_{i,j-1}(t) + \frac{t_{i+j}-t}{t_{i+j}-t_{i+1}} \cdot N_{i+1,j-1}, \quad j \in (1,k]; i \in [0,n] \tag{139}$$

$$\mathbf{T} = (t_0, t_1, \cdots, t_m), \quad t_{i+1} \geq t_i \text{ a vector of knots.}$$

(The convention $0/0 \equiv 0$ is assumed here for numerical purposes).

The formula (Eq. 139) defines the B-spline basis of order k ($p = k - 1$ degree). The knot vector $\mathbf{T}$ is an integral part of the definition. The knots t_0 and t_m are called the end knots and the other knots are called the interior knots.

The knot vector $\mathbf{T}$ forms the parametrization of the basis which influences its properties. In particular, the vector $\mathbf{T}$ decides on the basic classification of the B-spline bases:

a. Uniform and Non-uniform B-spline bases

b. Periodic and Non-periodic B-spline bases

Classification a pertains to the general rules of parametrization of a curve (see Section 1.2) and b concerns an ordering of knots in such a way as to get an open (non-periodic) or closed (periodic) curve as a result of the approximation. The uniform B-spline basis is said to be a standard spline basis in CAGD.

By the very definition, a B-spline basis of order k is a spline function of degree p ($p = k - 1$). This also means a pth degree of differentiability (smoothness) at the knots. The maximum differentiability at knots is ($k - 2$) and can deliberately be decreased (even to 0) by making some knots multiple knots. The sub-interval of the parameter axis: $[t_i, t_{i+1}]$ is called the ith span of the basis function. k successive spans form the ith support of the basis and so, for uniform splines with different knots, the length of each support is k. Fig. 44 shows some examples of B-spline basis functions.

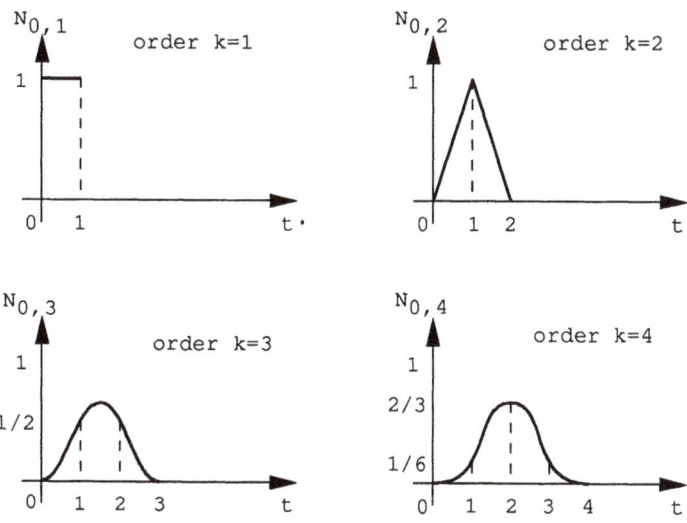

Fig. 44. B-spline basis functions $N_{0,k}(t)$ for $k = 1, 2, 3, 4$ [50]

The B-spline basis functions $N_{i,k}(t)$ have some important properties which influence the result of B-spline approximation and are the key to the behaviour of B-spline curves. The main properties of B-spline basis functions are summarized in Table 3.

Some additional properties of B-spline bases follow from the basis function definition (Eq. 139) and the properties listed in Table 3.

Table 3. Properties of the B-spline basis functions

Property	Analytical form
Local support	$N_{i,k}(t) \neq 0$ for $t \in [t_i, t_{i+k}]$ $N_{i,k}(t) = 0$ for $t \notin [t_i, t_{i+k}]$
Positivity	$N_{i,k} \geq 0$
Cauchy's relation (partition of unity)	$\sum_{i=0}^{n} N_{i,k}(t) = 1$
Translation (for uniform B-splines only)	$N_{i,k}(t) = N_{0,k}(t - i)$
Recursion	Given by the definition (Eq. 139)

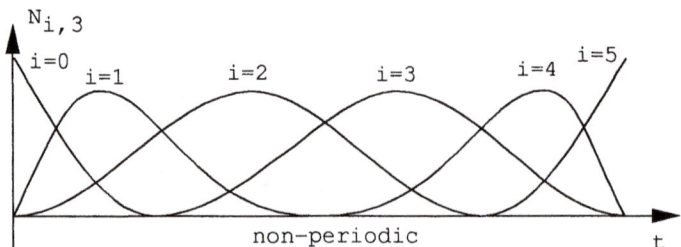

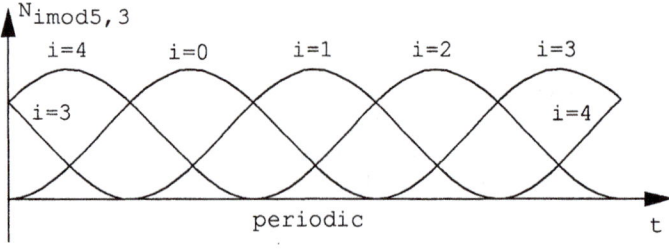

Fig. 45. Non-periodic and periodic basis functions [27]

o Multiplicity of Knots

The definition (Eq. 139) admits multiple knots. If $m_i \geq 0$ - multiplicity of an ith knot, then a multiple knot means that:

$$t_i = t_{i+1} = t_{i+2} = \cdots = t_{i+m_i-1}. \tag{140}$$

A knot vector $\mathbf{T}$ may contain identical knots up to k. The default multiplicity of all the knots is $m_i = 1$. In this case the B-spline basis is called a full spline basis. One effect on the basis function of a knot occurring with multiplicity $m_i > 1$ is to decrease the degree of differentiability of the basis function $N_{i,k}(t)$ at t_i from C^{k-2} to C^{k-m_i-1}. A basis arising from interior knots having multiplicity $m_i > 1$ is called a subspline basis. Multiple knots can lead to spans of zero length and, as a consequence, can reduce the corresponding width of the support of the basis function.

o Number of Nonzero Basis Functions

In the general case of multiple knots there are k nonzero basis functions between the knots and $k - m_i$ nonzero basis functions at the knots t_i with multiplicity m_i [83].

o Relation Between B-Spline and Bernstein Basis

There are two special cases when the B-spline basis reduces to the Bernstein basis:

- If $m_i = 0$ for all i, the B-spline basis is a spline basis of degree $k - 1$ and C^{k-1} differentiability everywhere, including at the knots. So the knots are actually pseudo-knots and the B-spline basis is a pseudo-spline and becomes the Bernstein basis [27].

- If a degenerate knot vector is set up, consisting only of the end points $t_0 = 0$ and $t_m = 1$, both with multiplicities $m_0 = m_m = k$:

$$\mathbf{T} = (\underbrace{0,0,\cdots,0}_{k},\underbrace{1,1,\cdots,1}_{k}) \tag{141}$$

then the B-spline basis is a degenerate basis and becomes the Bernstein basis. So, in both cases one gets

$$N_{i,k}(t) = \binom{k-1}{i} \cdot t^i \cdot (1-t)^{k-1-i}, \quad i = 0,1,\cdots,k-1. \tag{142}$$

The above property illustrates why B-splines are a proper spline extension of Bernstein polynomials.

3.8. General Form of B-Spline Curves

A B-spline curve is defined in the geometric form (Eq. 65), (Eq. 103) as a linear combination of polygon vertices $\mathbf{V}_i$, $i = 0, 1, \cdots, n$ and B-spline basis functions (Eq. 139). In this context, the vectors $\mathbf{V}_i$ are called de Boor points. The general form of a B-spline curve (function) is:

$$\mathbf{r}(t) = \sum_{i=0}^{s} \mathbf{V}_i \cdot N_{i,k}(t), \quad t \in [a, b] \tag{143}$$

where $\mathbf{T} = (t_0, t_1, \cdots, t_m)$ is a vector of knots. The subscripts s, m and the domain $[a, b]$ depend on the specific type of B-spline curve.

The curve $\mathbf{r}(t)$ consists of a set of points which can generally be calculated: $\mathbf{P}(t) = \mathbf{r}(t)$. In particular, the point $\mathbf{P}(t_i)$ on a curve, corresponding to the ith knot, is called the ith joint, and the arc of the curve between two successive joints $\mathbf{P}_i, \mathbf{P}_{i+1}$, corresponding to the span $[t_i, t_{i+1}]$, is called the ith segment of the curve. An example of Eq. 143 for one i^{th} segment of a B-cubic, uniform spline curve ($k = 4, p = 3$), after reparametrization $[t_i, t_{i+1}] \rightarrow [0, 1]$ is:

$$\mathbf{r}_i(t) = \sum_{j=1}^{i+3} \mathbf{V}_j \cdot N_{j,4}(t), \quad t \in [0, 1], \quad i = 0, 1, 2, \cdots \tag{144}$$

where the B-spline basis, expressed as the power basis (Eq. 57) is:

$$N_{j,4}(t) = \frac{1}{(4-1)!} \cdot \underbrace{\begin{bmatrix} 1 & -3 & 3 & -1 \\ 4 & 0 & -6 & 3 \\ 1 & 3 & 3 & -3 \\ 0 & 0 & 0 & 1 \end{bmatrix}}_{\mathbf{A}} \cdot \underbrace{\begin{bmatrix} 1 \\ t \\ t^2 \\ t^3 \end{bmatrix}}_{\mathbf{E}(t)} = \frac{1}{(4-1)!} \cdot \mathbf{A}(\mathbf{T}) \cdot \mathbf{E}(t) \tag{145}$$

where $\mathbf{T} = [t_0, t_1, t_2, t_3, t_4, t_5, t_6, t_7] = [-3, -2, -1, 0, 1, 2, 3, 4]$. One can compare it with Ferguson's (Eq. 81) and Bézier's (Eq. 112) cubic segments.

Eq. 144 can be generalized to express the kth order B-spline basis as the power basis (Eq. 57). For the uniform case one gets [64]:

$$N_{i,k}(t) = \frac{1}{(k-1)!} \cdot \mathbf{A}^k(\mathbf{T}) \cdot \mathbf{E}^k(t) \tag{146}$$

where the matrix $\mathbf{A}^k = [a_{ij}^k]$ is defined by the recursive formula:

$$\begin{cases} a_{ij}^k = a_{ij}^{k-1} - a_{i,j-1}^{k-1} + (j-1) \cdot a_{i-1,j}^{k-1} + (k-j+1) \cdot a_{i-1,j-1}^{k-1} \\ a_{0j}^{k-1} = a_{i0}^{k-1} = a_{kj}^{k-1} = a_{ik}^{k-1} = 0. \end{cases} \tag{147}$$

The specific types of multi-segment B-spline curves of kth order are given in Section 3.10. They depend on some important properties of B-spline curves.

3.9. Properties of B-Spline Curves

The properties of B-spline basis functions result in some important geometric properties of B-spline curves which make them a very useful tool in CAGD. Among those discussed below, some may be regarded as global and others as local properties. Some of those listed are identical or similar to those of Bézier curves. Many of them, however, are totally new and characterize B-spline curves exclusively.

o The Geometry Invariant Property (global)

The shape of a B-spline curve is determined only by its vertices and knot vector and is not related to the coordinate system. Moreover, Cauchy's relation (Table 3.) assures the shape of a curve to be invariant under coordinate transformation.

o The Variation Diminishing Property (global)

The number of intersections between an arbitrary plane and a B-spline curve is no more than the number of intersections between the plane and the curve defining polygon. Therefore, a B-spline curve assumes a shape that is a smoothed form of the curve defining polygon shape.

o The Order Independence Property (global)

The order k of the curve (the degree $p = k - 1$) and the number of polygon vertices ($N = n + 1$) may be chosen independently of each other. The only condition is $N > k$. This is an essential difference from Bézier's method where $k = N$.

o The Correspondence to Bézier Curves Property (global)

This follows directly from the multiplicity of knots property of the B-spline basis. If the number of vertices N equals the order of the curve k ($N = k$, without multiplicity of vertices) then a B-spline curve reduces to a Bézier curve. The knot vector then consists of k knots at either end and the curve has only one span (segment). The local deformability is lost in this case.

o The Convex Hull Property (local)

Since the B-spline basis functions (Eq. 139) are non-negative and sum to 1 (Cauchy's relation - see Table 3.), each point on a B-spline curve $P(t)$ is a convex combination of the polygon vertices V_i. This is as in the case of Bézier curves. However, there are only (at most) k vertices which determine a point P on the curve. This implies a much stronger convex hull property than is true for Bézier curves. Specifically, for a B-spline curve of degree $p = k - 1$, a given point lies within the convex hull of the k neighbouring vertices V_i. So, all points on a B-spline curve must lie within the union of all such convex hulls formed by taking k successive V_i. In Fig. 46, where the polygon is defined by vertices V_i, $i = 0, 1, \cdots, 10$, the shaded portions show how this region grows for $k = 2, 3, 4$ and $k \geq 5$, respectively.

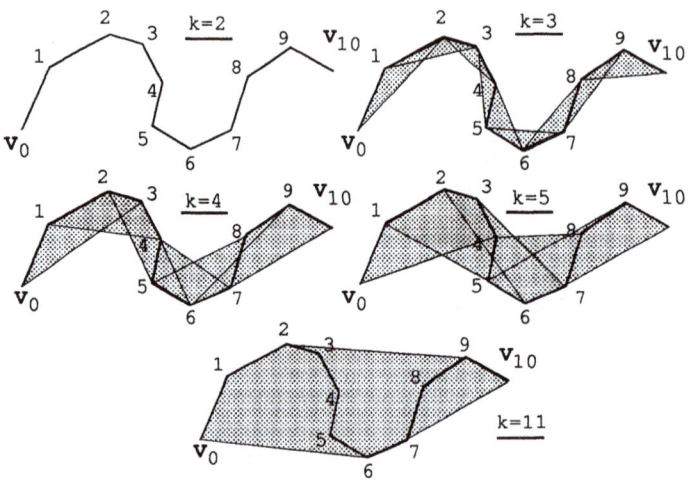

Fig. 46. The convex hull property [50]

o The Local Shape Control Property

An open B-Spline curve of order k ($p = k - 1$ degree), defined by $N = n + 1$ vertices $\mathbf{V}_i$ ($i = 0, 1, \cdots, n$) consists of $N_s = N - k + 1$ spans (segments) with $N_i = N - k$ interior knots. The shape of every segment is affected by k successive vertices (for example the segment i by the vertices $\mathbf{V}_i, \cdots, \mathbf{V}_{i+k-1}$). Conversely, no vertex influences more than k spans. This property permits local changes of the shape to be made, given an appropriate surplus of vertices ($N > k$), to some (but not all) segments of the curve. In other words, perturbing a single vertex of the polygon produces only a local perturbation of the curve in the vicinity of that vertex (Fig. 47). Note, that the Bézier curve does not possess this local property, which is desirable in CAGD.

o The End Points Property (local)

In general, a B-spline curve does not pass through the vertices $\mathbf{V}_i$ (unlike a Bézier curve which passes through the two end vertices). In particular, however, it is possible to force the B-spline curve to behave exactly like a Bézier curve. This is because of the multiple knots property. When there are k multiple knots at the two end knots (t_0 and t_m), then the corresponding end points of the defining polygon and the B-spline curve coincide, so: $\mathbf{V}_0 = \mathbf{P}(t_0), \mathbf{V}_n = \mathbf{P}(t_m)$. Moreover, the slopes of the curve at those points are the directions of the vectors: $(\mathbf{V}_1 - \mathbf{V}_0)$ and $(\mathbf{V}_n - \mathbf{V}_{n-1})$.

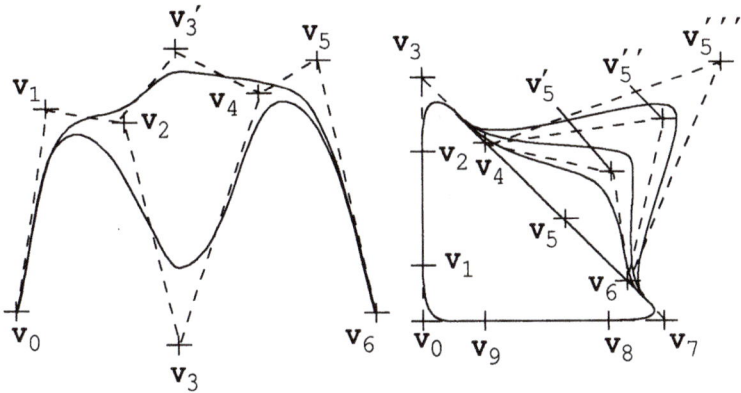

Fig. 47. The control shape property

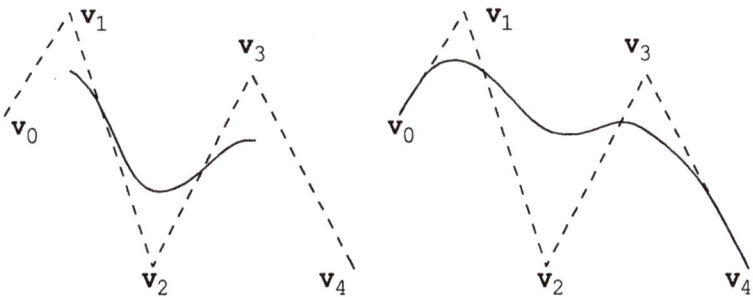

Fig. 48. The end points property

o The Multiple Vertices Property (local)

The multiplicity of the ith vertex means that m_i successive vertices coincide:

$$\mathbf{V}_i = \mathbf{V}_{i+1} = \mathbf{V}_{i+2} = \cdots = \mathbf{V}_{i+m_i-1} \qquad (148)$$

The effect of a vertex multiplicity depends on the degree of multiplicity m_i [23]. In CAGD, the most important cases are:

- $m_i = k$ (the order of the B-spline curve)
    In this case, the convex hull of the set (Eq. 148) as well as the ith segment

of the B-spline curve determined by these vertices reduces to the vertex $\mathbf{V}_i$ itself. As a consequence, the curve must pass through the point $\mathbf{V}_i$.

- $m_i = k - 1$ (the degree of the B-spline curve)

  In this case, the ith segment of the B-spline curve becomes a linear section the length of which is

$$l_i = \frac{1}{(k-1)!} \cdot \mathbf{V}_i - \mathbf{V}_{i+1}. \tag{149}$$

  Using this property twice, one gets two successive segments, linear, with a sharp corner between them.

- $m_i = k - 2$ (the degree of differentiability of B-spline curve)

  In this case, the ith segment of the B-spline curve is tangent to the edges of polygon at the two end points of the segment.

The three cases mentioned are illustrated in Fig. 49.

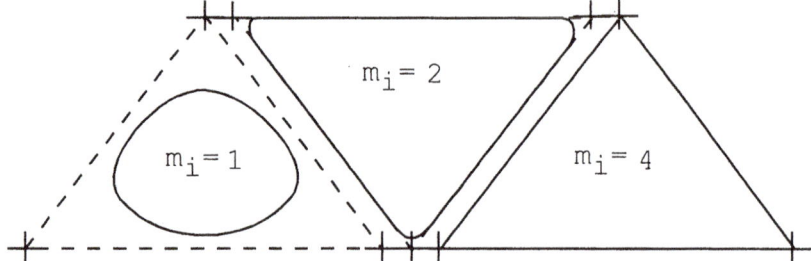

Fig. 49. The multiple vertices property (k=4)

o The Collinear Vertices Property (local)

This property follows from two other properties: the localness of B-spline curve approximation schemes and the variation diminishing property. The former says that k successive vertices fully determine one segment of the B-spline curve. The latter assures the B-spline approximation reproduces straight lines. The conjunction of the two indicates that the segment determined by $(k-1)$ collinear vertices is also linear (Fig. 50a) while with k collinear vertices the curve becomes tangent to the polygon (Fig. 50b).

o The Discontinuities Property (local)

This property follows from the property of the B-spline basis. Using multiple knots, all kinds of discontinuities (knuckles, curvature jumps etc.) are easily obtainable; a knot of multiplicity m_i reduces differentiability of the curve at this knot by $(m_i - 1)$ orders.

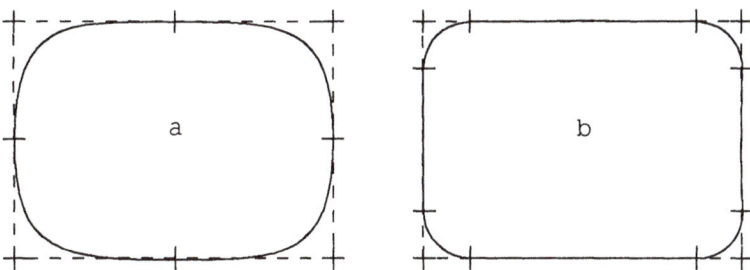

Fig. 50. The collinear vertices property (k=4)

3.10. Particular Forms of B-Spline Curves

The general form of a B-spline curve was given in Section 3.8 as

$$r(t) = \sum_{i=0}^{s} V_i \cdot N_{i,k}(t), \quad t \in [a, b] \tag{150}$$

where $T = (t_0, t_1, \cdots, t_n)$ is a vector of knots.

Now the particular forms for different types of B-spline curves will be given which describe the details of the definition (Eq. 139). According to the classification, B-spline curves fall into the three categories:

- Uniform - non-uniform B-spline curves,

- Open - closed B-spline curves,

- B-spline curves without the coincidence between the end points of the polygon and the curve versus B-spline curves with the coincidence of these points.

Of several types of B-spline curves listed above, the uniform B-spline curves possessing the end points coincidence property (the same property as in the case of Bézier curves) are regarded as standard B-spline curves in CAGD.

Formulae for different B-spline curves are assembled in Tables 4. and 5.

Table 4. Particular forms of uniform B-spline curves

Open curve, without the coincidence of the end points
$$\mathbf{r}(t) = \sum_{i=0}^{n} \mathbf{V}_i \cdot N_{i,k}(t) \quad t \in [0, n-k+2]$$ $$\mathbf{T} = (t_0, t_1, \cdots, t_{n+k})$$ $$t_i = i - k + 1, \quad i = 0, 1, \cdots, n+k$$ $$(151)$$
Open curve, with the coincidence of the end points
$$\mathbf{r}(t) = \sum_{i=0}^{n} \mathbf{V}_i \cdot N_{i,k}(t) \quad t \in [0, n-k+2]$$ $$\mathbf{T} = (t_0, t_1, \cdots, t_{n+k})$$ $$t_i = 0, \quad i = 0, 1, \cdots, k-1$$ $$t_{i+k} = i + 1, \quad i = 0, 1, \cdots, n-k$$ $$t_{i+n+1} = n - k + 2, \quad i = 0, 1, \cdots, k-1$$ $$(152)$$
Closed curve
$$\mathbf{r}(t) = \sum_{i=0}^{n+k-1} \mathbf{V}_{i \bmod (n+1)} \cdot N_{i,k}(t) \quad t \in [0, n+1]$$ $$\mathbf{T} = (t_0, t_1, \cdots, t_{n+2k-1})$$ $$t_i = i - k + 1, \quad i = 0, 1, \cdots, k-2$$ $$t_{i+k-1} = i, \quad i = 0, 1, \cdots, n+1$$ $$t_{i+n+k+1} = i + n + 2, \quad i = 0, 1, \cdots, k-2$$ $$(153)$$

Table 5. Particular forms of non-uniform B-spline curves

Open curve, without the coincidence of the end points
$$\mathbf{r}(t) = \sum_{i=0}^{n} \mathbf{V}_i \cdot N_{i,k}(t) \quad t \in [a_{k-1}, a_{k+1}]$$ $$\mathbf{T} = (t_0, t_1, \cdots, t_{n+k})$$ $$t_i = a_i, \quad i = 0, 1, \cdots, n+k \tag{154}$$
Open curve, with the coincidence of the end points
$$\mathbf{r}(t) = \sum_{i=0}^{n} \mathbf{V}_i \cdot N_{i,k}(t) \quad t \in [a_0, a_{n-k+2}]$$ $$\mathbf{T} = (t_0, t_1, \cdots, t_{n+k})$$ $$t_i = a_0, \quad i = 0, 1, \cdots, k-1$$ $$t_{i+k} = a_{i+1}, \quad i = 0, 1, \cdots, n-k$$ $$t_{i+n+1} = a_{n-k+2}, \quad i = 0, 1, \cdots, k-1 \tag{155}$$
Closed curve
$$\mathbf{r}(t) = \sum_{i=0}^{n+k-1} \mathbf{V}_{i \bmod (n+1)} \cdot N_{i,k}(t) \quad t \in [a_0, a_{n+1}]$$ $$\mathbf{T} = (t_0, t_1, \cdots, t_{n+2k-1})$$ $$t_i = a_0 - (a_{n+1} - a_{i+n-k+2}), \quad i = 0, 1, \cdots, k-2$$ $$t_{i+k-1} = a_i, \quad i = 0, 1, \cdots, n+1$$ $$t_{i+n+k+1} = a_{n+1} + (a_{i+1} - a_0), \quad i = 0, 1, \cdots, k-2 \tag{156}$$

3.11. Algorithms of B-Spline Curves

The algorithms of a B-spline curve serve for manipulating the shape in the process of geometric modelling of the curve.

o Insertion of Knots Algorithm

The algorithm can be applied in the case of non-uniform B-spline curves. A new, additional knot can be inserted between existing knots to increase the number of polygon vertices and the number of segments of a curve while maintaining its shape. If

$$\mathbf{T} = [t_0, t_1, \cdots, t_i, t_{i+1}, \cdots, t_m] \tag{157}$$

is an existing knot vector and a new knot t_a, to be inserted, falls into the interval $(t_i, t_{i+1}]$, so $t_i < t_a \leq t_{i+1}$, then the new knot vector (before and after renumbering) is:

$$\overline{\mathbf{T}} = [t_0, t_1, \cdots, t_i, t_a, t_{i+1}, \cdots, t_m] = [\bar{t}_0, \bar{t}_1, \cdots, \bar{t}_{m+1}] \tag{158}$$

and the new polygon vertex can be calculated [23] as follows:

$$\overline{\mathbf{V}}_j = (1 - \lambda_j) \cdot \mathbf{V}_{j-1} + \lambda_j \cdot \mathbf{V}_j \tag{159}$$

where

$$\lambda_j = \begin{cases} 1 & \text{for } j \leq i - k + 1 \\ \dfrac{t_a - t_j}{t_{j+k-1} - t_j} & \text{for } j \in [i - k + 2, i] \\ 0 & \text{for } j \geq i + 1. \end{cases} \tag{160}$$

As a consequence, the two definitions of existing spline curves are:

$$\mathbf{r}(t) = \sum_i \mathbf{V}_i \cdot N_{i,k}(t) \equiv \sum_j \overline{\mathbf{V}}_j \cdot N_{j,k}(t) = \overline{\mathbf{r}}(t). \tag{161}$$

o The Positive Algorithm

The algorithm serves for the calculation of $\mathbf{r}(t)$ at a given value $t = t_s$. In the case of B-spline curves, a standard positive algorithm is the deBoor algorithm. It is very similar to that of the de Casteljau algorithm for Bézier curves (Eq. 131) and follows from the recursion property of the B-spline basis (Eq. 139). In the deBoor algorithm a set of auxiliary vertices $\mathbf{V}_{jl}$ is recursively calculated which converges to a point on a B-spline curve.

Let the given value t_s falls in the sub-interval $[t_i, t_{i+1})$, so the relation holds: $t_i \leq t_s < t_{i+1}$, $i = 0, 1, \cdots, r$, where:

$$r = \begin{cases} n + k & \text{for an open curve} \\ n + 2k - 1 & \text{for a closed curve.} \end{cases} \tag{162}$$

Then the subscripts:
$$m = i - k + 1 \tag{163}$$
describe polygon vertices which define a segment of a B-spline curve containing the point on the curve, to be found: $\mathbf{P}_s = \mathbf{r}(t_s)$. The deBoor algorithm is as follows:

$$\mathbf{V}_{j,0}(t_s) = \begin{cases} \mathbf{V}_j & \text{for an open curve} \\ \mathbf{V}_{jmod(n+1)} & \text{for a closed curve} \end{cases} \tag{164}$$

for $j = m, m+1, m+2, \cdots, m+k-1$, and

$$\mathbf{V}_{j,1}(t_s) = (1-\lambda) \cdot \mathbf{V}_{j-1,l-1}(t_s) + \lambda \cdot \mathbf{V}_{j,l-1}(t_s) \tag{165}$$

where
$$\lambda = \frac{t_s - t_j}{t_{j+k-1} - t_j}, \quad \begin{array}{l} l = 1, 2, \cdots, k-1 \\ j = m+l, m+l+1, \cdots, m+k-1. \end{array} \tag{166}$$

It can be proven [23], that $\mathbf{P}_s = \mathbf{r}(t_s) = \mathbf{V}_{i,k-1}(t_s)$. A plotting procedure (Fig. 51) illustrates the deBoor algorithm.

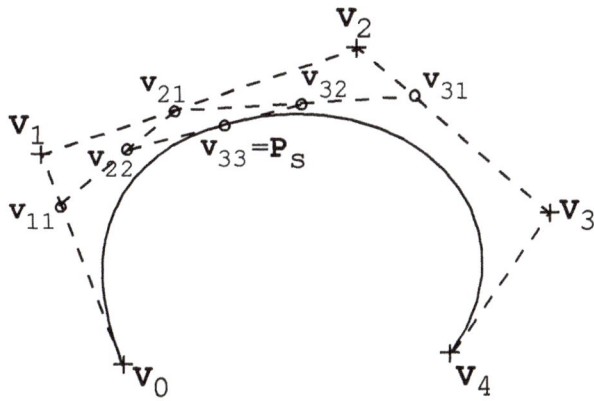

Fig. 51. Interpretation of the deBoor algorithm

o The Inverse Algorithm

The algorithm for B-spline curves serves for solving an interpolation problem by an approximation technique. The B-spline approach (Eq. 143) defines a curve $\mathbf{r}(t)$ that only approximates the given (control) points $\mathbf{V}_i$. As a result, a sequence of points $\mathbf{P}_j = \mathbf{r}(t_j)$ on the curve is obtained. An inverse problem is:

to find a set of vertices $\mathbf{V}_i$ and then a B-spline curve with a set of points $\mathbf{P}_j$ on the curve given. This is, in fact, an interpolation problem.

The inverse algorithm for uniform cubic B-spline curves will be presented which is an alternative approach to that discussed in Section 2.4. A cubic spline interpolation approach was based on the continuity conditions imposed on a Ferguson cubic curve. The inverse algorithm approach for B-cubic splines is based on the properties of uniform B-spline basis functions $N_{i,k}(t)$, for $k = 4$. Both of them are totally equivalent.

$$\text{Let } \mathbf{P}_j \quad \begin{cases} j = 1, 2, \cdots, n - 1 & \text{- for an open curve} \\ j = 0, 1,, \cdots, n & \text{- for a closed curve} \end{cases} \tag{167}$$

be a set of given points on a curve to be found. The inverse algorithm problem is one of finding vertices (control points) of a defining polygon:

$$\mathbf{V}_i \quad \begin{cases} i = 0, 1, \cdots, n & \text{- for an open curve} \\ i = -1, 0, 1, \cdots, n + 1 & \text{- for a closed curve.} \end{cases} \tag{168}$$

Due to the properties of B-cubic spline basis functions $N_{i,4}(t)$ (Eq. 139) for both open and closed curves the following system of linear equations holds:

$$\mathbf{P}_m = \frac{1}{6} \cdot \mathbf{V}_{m-1} + \frac{2}{3} \cdot \mathbf{V}_m + \frac{1}{6} \cdot \mathbf{V}_{m+1} \tag{169}$$

$m = 1, 2, \cdots, n - 1$ for an open curve, and $m = 0, 1, \cdots, n$ for a closed curve.

Since there are two less equations than unknowns in both cases, the following additional conditions are applied to open and closed curves, respectively:

$$\mathbf{V}_0 = \mathbf{V}_1, \quad \mathbf{V}_n = \mathbf{V}_{n-1} \text{ - for an open curve} \tag{170}$$
$$\mathbf{V}_{-1} = \mathbf{V}_n, \quad \mathbf{V}_0 = \mathbf{V}_{n+1} \text{ - for a closed curve.} \tag{171}$$

The conditions of Eq. 170 state that the curvature is 0 at both ends of the open curve (so, the curve is to be a natural spline curve - compare with Eq. 88, Eq. 101 for the interpolation of a cubic spline). These conditions simplify the inverse algorithm to the same form as that for a closed curve. The conditions of Eq. 171 are those for joining the starting and end points with continuity up to the curvature vector to form a closed, smooth curve.

Substituting Eq. 170 (for an open curve) or Eq. 171 (for a closed curve) into Eq. 169, one gets the resulting systems of linear equations expressed in matrix form:

- Open curve:
$$\mathbf{V} \cdot \mathbf{N} = \mathbf{P} \tag{172}$$

where

$$\mathbf{P} = [\mathbf{P}_1, \mathbf{P}_2, \cdots, \mathbf{P}_{n-1}] \tag{173}$$
$$\mathbf{V} = [\mathbf{V}_1, \mathbf{V}_2, \cdots, \mathbf{V}_{n-1}] \tag{174}$$

and

$$\mathbf{N} = \frac{1}{6} \begin{bmatrix} 5 & 1 & 0 & 0 & 0 & \cdots & 0 & 0 & 0 & 0 & 0 \\ 1 & 4 & 1 & 0 & 0 & \cdots & 0 & 0 & 0 & 0 & 0 \\ 0 & 1 & 4 & 1 & 0 & \cdots & 0 & 0 & 0 & 0 & 0 \\ \cdots & \cdots & \cdots & \cdots & \cdots & \cdots & \cdots & \cdots & \cdots & \cdots & \cdots \\ 0 & 0 & 0 & 0 & 0 & \cdots & 1 & 4 & 1 & 0 & 0 \\ 0 & 0 & 0 & 0 & 0 & \cdots & 0 & 0 & 0 & 1 & 5 \end{bmatrix}^T_{(n-1)\cdot(n-1)} \tag{175}$$

- Closed curve:
$$\overline{\mathbf{V}} \cdot \overline{\mathbf{N}} = \overline{\mathbf{P}} \tag{176}$$

where

$$\overline{\mathbf{P}} = [\mathbf{P}_1, \mathbf{P}_2, \cdots, \mathbf{P}_n] \tag{177}$$
$$\overline{\mathbf{V}} = [\mathbf{V}_1, \mathbf{V}_2, \cdots, \mathbf{V}_n] \tag{178}$$

and

$$\overline{\mathbf{N}} = \frac{1}{6} \begin{bmatrix} 4 & 1 & 0 & 0 & 0 & \cdots & 0 & 0 & 0 & 0 & 1 \\ 1 & 4 & 1 & 0 & 0 & \cdots & 0 & 0 & 0 & 0 & 0 \\ 0 & 1 & 4 & 1 & 0 & \cdots & 0 & 0 & 0 & 0 & 0 \\ \cdots & \cdots & \cdots & \cdots & \cdots & \cdots & \cdots & \cdots & \cdots & \cdots & \cdots \\ 0 & 0 & 0 & 0 & 0 & \cdots & 1 & 4 & 1 & 0 & 0 \\ 1 & 0 & 0 & 0 & 0 & \cdots & 0 & 0 & 0 & 1 & 4 \end{bmatrix}^T_{(n+1)\cdot(n+1)} \tag{179}$$

It can be proven that the matrices $\mathbf{N}$ and $\overline{\mathbf{N}}$ (both predominantly three-diagonal ones) are nonsingular, and so $\det\mathbf{N} \neq 0$ and $\det\overline{\mathbf{N}} \neq 0$. Therefore, both systems (Eq. 172 and Eq. 176) have unique solutions:

- Open curve:
$$\mathbf{V} = \mathbf{P} \cdot \mathbf{N}^{-1}, \quad \mathbf{V}_0 = \mathbf{V}_1, \quad \mathbf{V}_n = \mathbf{V}_{n-1} \tag{180}$$

- Closed curve:
$$\overline{\mathbf{V}} = \overline{\mathbf{P}} \cdot \overline{\mathbf{N}}^{-1}, \quad \mathbf{V}_{-1} = \mathbf{V}_n, \quad \mathbf{V}_{n+1} = \mathbf{V}_0. \tag{181}$$

Examples of B-spline curves are shown in Figs. 52, 53 and 54.

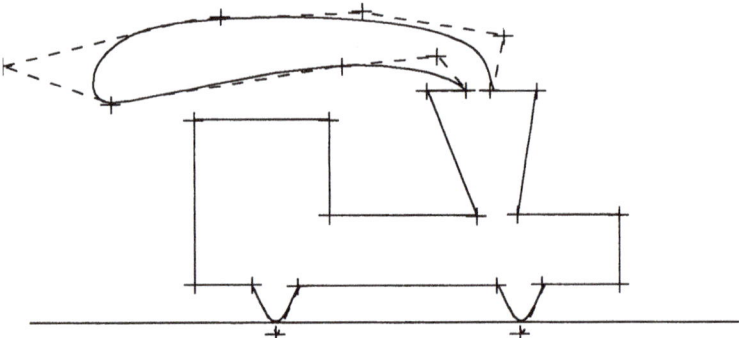

Fig. 52. Example of a B-spline curve (1)

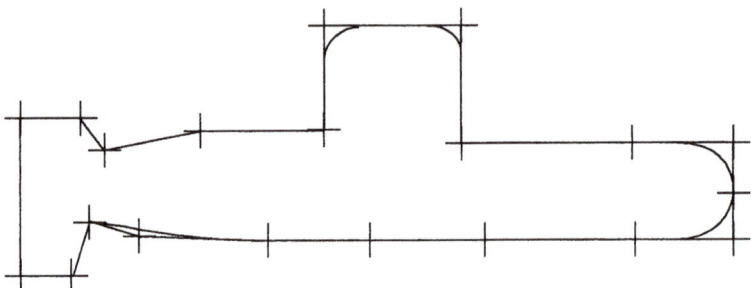

Fig. 53. Example of a B-spline curve (2)

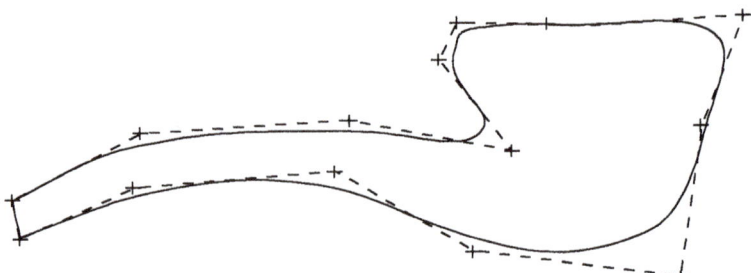

Fig. 54. Example of a B-spline curve (3)

3.12. Rational B-Spline Approximation (NURBS)

As has been mentioned earlier, simple geometric shapes, such as circles, ellipses and parabolas were used in medieval ship design. These curves are commonly represented using implicit equations. We have already seen that using the implicit form, free-form curves and surfaces cannot easily be modelled. This is one of the reasons why parametric forms such as Bézier and B-spline curves and the PDE method were introduced, with the most common form used in the computational representation of these curves being parametric polynomial forms.

In present day ship design, the ship is represented by families of curves collectively called the ship lines plan, consisting primarily of planar curves known as the waterlines, sections and buttock lines. B-spline curve representation is one of the most suited representations for ship form design. Unfortunately, however, it is not suited for the precise representation of circles, conics and other primitives which we may need. Therefore, it would be of benefit if a geometric modeller could be provided which had a scope which encompassed both free-form shapes as well as the representation of well-known geometric primitives such as primitives.

The major disadvantage, according to Tiller [75], of a system which incorporates various types of curve description (implicit and free-form, say) is that the computer code required for this representation grows much more rapidly than for one based on a single type for all geometric entities. For this reason, it is useful to generate all curves and surfaces in the same mathematical form. It was with this in mind that the idea of Non Uniform Rational B-Splines (or as they are referred to, NURBS) were introduced.

Non uniform rational B-splines were extended from B-splines by Versprille in 1975 [77]. The term rational refers to the ratio of polynomials that characterizes this approach, i.e. a NURBS curve is the ratio of two B-spline curves; and similarly for surfaces. To define a rational B-spline curve we make use of homogeneous coordinates as will be illustrated.

3.13. Homogeneous Coordinates and NURBS

Homogeneous coordinates are used to represent points in N dimensional space in terms of points in $(N+1)$ dimensional space. In particular, we can use homogeneous coordinates to represent points in $3D$ space in terms of $4D$ space coordinates. Alternatively, we can say that the projective N space is embedded in Euclidean $(N+1)$ dimensional space [26].

For example, when producing on a computer screen images of $3D$ objects, $2D$ projections are required. If we project a $3D$ curve, onto the flat plane $w =$const, and the projection lines converge to a common point O (known as the viewpoint), then we say we have a central projection or perspective view of the object. Thus, in Fig. 55 the plane $w =$const is termed the projection plane and is embedded in Euclidean $3D$ space.

If we now denote the coordinate axes of the Euclidean space by x, y and w, and let X, Y be the coordinate system of the projection plane parallel to x, y and with origin $(x, y, w) = (0, 0, 1)$, then any point $\mathbf{V}'$ in the projection plane will determine a line OV' which can be defined by any point $\mathbf{V}$ along this line. The coordinates (wx, wy, w) of $\mathbf{V}$ are called the homogeneous coordinates of $\mathbf{V}'$. Thus, for any point in $3D$ space, $\mathbf{V}$ is embedded along the projection line OV at the point $\mathbf{V}'$ where OV meets the projection plane.

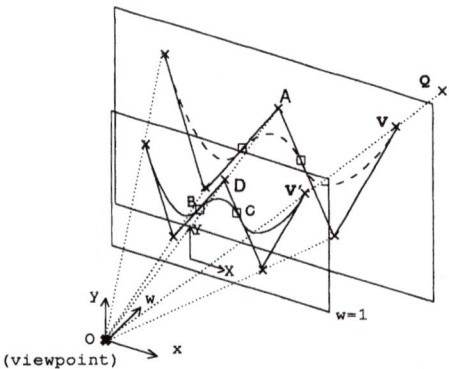

Fig. 55. The rational B-spline curve projected from an integral curve

The position of $\mathbf{V}$ along OV' is arbitrary, that is if $\mathbf{V}$ and $\mathbf{Q}$ are two different points along OV', their coordinates are both the homogeneous coordinates of $\mathbf{V}'$. Additionally, if $\mathbf{V}$ lies in the plane $x, y, w = 0$ then OV will never intersect the X, Y plane and hence $\mathbf{V}$ is called a point at infinity in the direction x, y.

Thus, each point in the x, y, w coordinate system can be mapped onto the X, Y plane with the perspective map

$$H\{(wx, wy, w)\} = \begin{cases} (wx/w, wy/w) & \text{if } w \neq 0 \\ \begin{array}{l}\text{point at infinity on} \\ \text{the line from the} \\ \text{origin through } (x, y).\end{array} & \text{if } w = 0 \end{cases} \tag{182}$$

Using this notation, the rational B-spline in $3D$ space can be represented in terms of $4D$ homogeneous coordinates simply by adding an extra z coordinate.

Thus, if the B-spline is given in homogeneous space by the vector equation

$$\mathbf{r}^W(t) = \sum_{i=0}^{n} \mathbf{V}_i^W \cdot N_{i,k}(t) \tag{183}$$

in which the $\mathbf{V}_i^W$ are the set of 4D control points defined by $\mathbf{V}_i^W = (w_i x_i, w_i y_i, w_i z_i, w_i)$ for $i = 0, \cdots, n$, and the control points in 3D space are defined by $\mathbf{V}_i = H\{\mathbf{V}_i^W\} =$

(x_i, y_i, z_i), then the rational B-spline curve will be given by

$$\mathbf{r}(t) = H\{\mathbf{r}^W(t)\} = H\left\{\sum_{i=0}^{n} \mathbf{V}_i^W \cdot N_{i,k}(t)\right\} = \frac{\sum_{i=0}^{n} \mathbf{V}_i N_{i,k}(t) w_i}{\sum_{i=0}^{n} N_{i,k}(t) w_i} = \sum_{i=0}^{n} \mathbf{V}_i \cdot R_{i,k}(t) \quad (184)$$

where

$$R_{i,k}(t) = \frac{N_{i,k}(t) \cdot w_i}{\sum_{j=0}^{n} N_{j,k}(t) \cdot w_j}. \quad (185)$$

The coordinates w_i are termed the weights of the control points and each of them are associated with a distinct control point [59].

3.14. Influence of Weights

To illustrate how we can obtain a rational curve from the projection of a B-spline curve, and to examine the influence of the weights, w_i, we follow the approach of Barsky [2].

In this example we consider the case of a rational curve lying on a 2D plane, given as $w = 1$. On the plane the curve will be defined by the coordinates (x, y), and can be thought of as being the projection of a curve in 3D space onto this plane. The original curve in Euclidean 3-space will therefore be represented by the homogeneous coordinates (wx, wy, w). The B-spline curve and the rational B-spline are illustrated in Fig. 55 in which the B-spline curve is displayed with a dashed line, and the rational B-spline as a solid line.

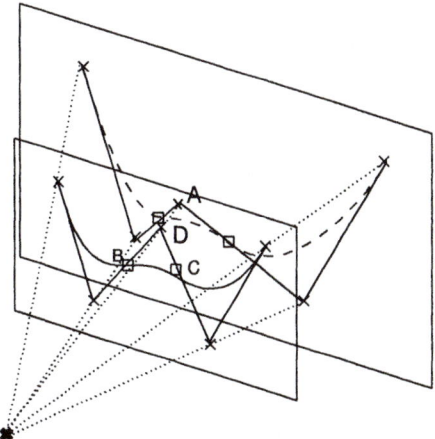

Fig. 56. Control point A moved towards the centre of projection

Consider the effect of moving one of the homogeneous control points (wx_i, wy_i, w_i). If we consider control point **A** in Fig. 55 then this forms a projection line OA to the viewpoint **O**. The corresponding NURBS control point D lies on the projection plane $w = 1$. If we now move **A** along this projection line towards the viewpoint, then the NURBS control point D will remain fixed. However, the local section of the rational curve BC, will move downwards and away from **D**, as in Fig. 56. In the limiting case, as the homogeneous control point approaches the viewpoint, the rational curve segment, BC, will tend to a straight line connecting the control points either side of control point **D**. This then corresponds to a weight w_i which is zero, since the control point **D** has no influence on the section of curve.

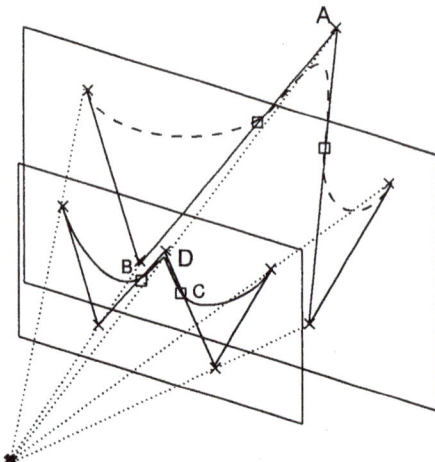

Fig. 57. Control point A moved away from the centre of projection

If we now move the homogeneous control point **A** away from the viewpoint, again along OA, as in Fig. 57, the rational control point **D** remains fixed. However, the local section of curve BC will now move towards the rational control point **D**. Again, in the limiting case, as the homogeneous control point **A** moves towards infinity along OA, the rational curve section BC becomes more pronounced and tends to control point **D**. The control point is said to have a weight w_i of infinity implying that the rational curve interpolates point **D**.

To summarize, the weights, w_i, are such that they only alter the section of the curve $[t_i, t_{i+k+1})$, which means that they offer a local control of the curve or surface. Furthermore, w_i can have a value ranging from zero (the control point does not influence the curve) to infinity (the curve interpolates the control point). We therefore see how shape modification is possible using weights: as w_i increases, the curve is pulled towards its control point V_i, as w_i decreases the curve moves away from V_i, and so the weights act as a push/pull movement of the rational curve, as can be seen in Fig. 58 for different values of the weight w_1.

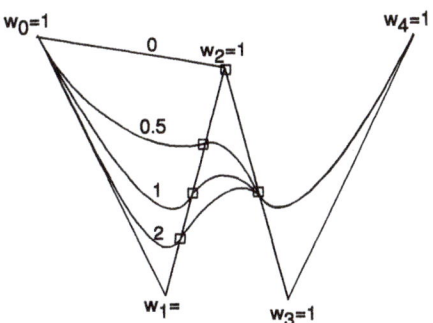

Fig. 58. Different weighted spline curves

3.15. Rational Basis Function and Curve Properties

NURBS have many properties in common with B-splines (see Section 3.9). In fact, if we take $w_i = 1$ for all i, and using Cauchy's relation (Table 3.), then substituting into Eq. 184, we obtain the formulation of a B-spline curve. Furthermore, taking the knot vector $\mathbf{T} = (0, \cdots, 0, 1, \cdots, 1)$ with multiplicity k we obtain rational and integral Bézier curves. However, there are some properties of NURBS which are totally new and characterize NURBS curves exclusively:

o The rational basis functions form a partition of unity, i.e.

$$\sum_{i=0}^{n} R_{i,k}(t) = 1 \quad \text{for all } t \in [t_0, t_m]. \tag{186}$$

o The shape of a NURBS curve is invariant under linear affine transformations (as are B-spline curves). However, unlike B-spline curves, NURBS curves possess the property of being unaffected by perspective projections. This means that the projection of the B-spline curve defined in homogeneous coordinate space is equivalent to the curve which results from the same projection applied to the homogeneous control vertices from which a curve is then obtained.

o From Eq. 186, it also follows that each of the rational basis functions $R_{i,k}(t)$ must be non-negative, since the weights $w_i > 0$.

o If a control point is moved or a weight changed it will only affect the curve in $k+1$ knot spans.

Furthermore since NURBS possess many of the same properties as B-splines, algorithms already designed for B-spline manipulation can be used to design curves using NURBS (see Section 3.11). However along with the usual methods of moving the control points or changing the knot vector, we have an extra degree of freedom in that the weights can alter the shape of the curve with a push/pull action. Furthermore, it is the important property of being able to represent conics and other primitives which is the biggest advantage of NURBS over conventional splines, and it is this which we will now illustrate.

3.16. Parametrization of a Conic

Consider the segment $P_0P'P_2$ of the conic section shown in Fig. 59. In order to describe the segment we construct the tangents at the points $\mathbf{P}_0, \mathbf{P}', \mathbf{P}_2$. These intersect at $\mathbf{P}_1, \mathbf{A}, \mathbf{B}$. By regarding $\mathbf{P}_0, \mathbf{P}_1, \mathbf{P}_2$ as fixed, then $\mathbf{A}$ and $\mathbf{B}$ will vary as $\mathbf{P}'$ moves along the conic. The ratios $g_1 = \mathbf{P}_1\mathbf{A}/\mathbf{P}_0\mathbf{A}$ and $g_2 = \mathbf{P}_1\mathbf{B}/\mathbf{P}_2\mathbf{B}$ depend on $\mathbf{P}'$ and so we can provide ourselves with a parametrization of the conic. From Faux

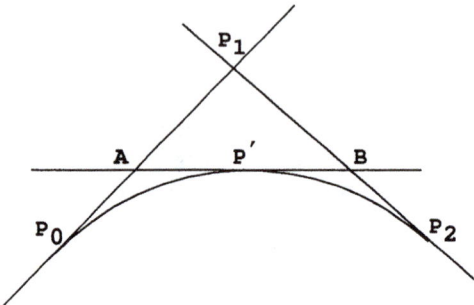

Fig. 59. A segment of the conic section

and Pratt [26], the point $\mathbf{P}'$ has position vector given by

$$\mathbf{P}' = \frac{g_1\mathbf{P}_0 + 2\mathbf{P}_1 + g_2\mathbf{P}_2}{g_1 + 2 + g_2}. \tag{187}$$

By letting $g_1 = w_0(1-u)/w_1u$ and $g_2 = w_2u/w_1(1-u)$ be in terms of a parameter u and where w_0, w_1 and w_2 are constants known as weights we obtain

$$\mathbf{P}' = \frac{w_0\mathbf{P}_0(1-u)^2 + 2w_1\mathbf{P}_1u(1-u) + w_2\mathbf{P}_2u^2}{w_0(1-u)^2 + 2w_1u(1-u) + w_2u^2} \tag{188}$$

which is a rational quadratic parametric equation for the conic section with tangents $\mathbf{P}_0\mathbf{P}_1$ and $\mathbf{P}_1\mathbf{P}_2$ and having $g_1g_2 = k^2 = w_0w_2/w_1^2$ where k^2 is called the conic shape factor [26].

The conic shape factor k^2 is constant for any conic and so the parametrization of the curve can be altered by changing the weights whilst keeping the ratio w_0w_2/w_1^2 constant. In fact, if we fix the end weights, $w_0 = w_2 = 1$ say, we can obtain the complete family of conics by adjusting w_1: hence, if $k^2 < 1$ we obtain an ellipse, if $k^2 = 1$ a parabola and $k^2 > 1$ gives a hyperbola.

Thus, we have a general formulation for the rational parametrization of a conic section. If we now consider the class of NURBS curves we shall see how we can represent these conics in terms of NURBS.

3.17. NURBS Representation of Conics

Since Eq. 188 is a quadratic we can work with the class of quadratic NURBS

$$r(t) = \sum_{i=0}^{2} V_i \cdot R_{i,k}(t) \tag{189}$$

where the rational basis functions are defined over the knot vector $T = (0,0,0,1,1,1)$. Thus, from Eq. 189 the equation of the quadratic NURBS is given by

$$r(t) = \frac{w_0 V_0 (1-t)^2 + 2 w_1 V_1 t(1-t) + w_2 V_2 t^2}{w_0 (1-t)^2 + 2 w_1 t(1-t) + w_2 t^2} \tag{190}$$

which is exactly of the form of the conics given by Eq. 188. As previously, we can alter the NURBS representation of the conic with its weights, w_i.

Alternatively, we can select a conic by specifying a third point on it, associated with some parameter, $t = 1/2$, say, and by fixing the end weights $w_0 = w_2 = 1$ [60]. Substitution into Eq. 190 yields

$$S = \frac{1}{1+w_1} M + \frac{w_1}{1+w_1} V_1 \tag{191}$$

where M is the midpoint of the chord $V_0 V_2$. If we introduce a parameter s, such that $w_1 = s/(1-s)$ then we obtain

$$S = (1-s)M + sV_1 \tag{192}$$

which now gives us a new shape design tool that gives a linear interpolation between M and V_1.

Thus, if $s = 0$, we get a line segment, if $0 < s < 1/2$ an ellipse, $s = 1/2$ yields a parabola and for $1/2 < s < 1$ a hyperbola is formed as in Fig. 60. Setting $s = 1$ in Eq. 192 interpolates the control point V_1 and we can see that this gives an infinite weight w_1 which is consistent with the properties of weights.

To generate a circular arc we need the triangle $V_0 V_1 V_2$ to be isosceles (since symmetry is required for the arc). As $w_1 = s/(1-s) = MS/SV_1$ we obtain the relationship $w_1 = \cos \theta$, where θ is the angle $\angle V_1 V_0 M$ (see Fig. 61).

Thus, to construct a circle we generate four 90° arcs and piece them together. The circle will then be defined as

$$r(t) = \sum_{i=0}^{8} V_i \cdot R_{i,3}(t) \tag{193}$$

where $(w_0, w_1, \cdots, w_8) = (1, \sqrt{2}/2, 1, \sqrt{2}/2, 1, \sqrt{2}/2, 1, \sqrt{2}/2, 1)$ (since $\theta = 45°$) with the knot vector $T = (0,0,0,1/4,1/4,1/2,1/2,3/4,3/4,1,1,1)$ and the V_i's forming the circumscribing square of the circle, as in Fig. 62.

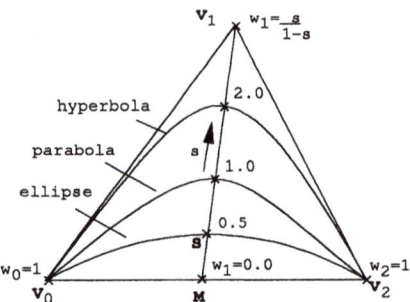

Fig. 60. The family of conics generated for various s

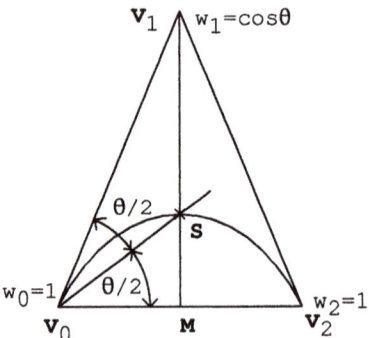

Fig. 61. Rational representation of a circular arc

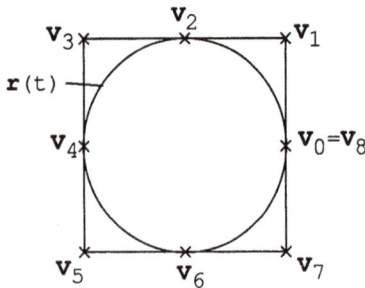

Fig. 62. NURBS representation of the circle

CHAPTER 4
CURVE GENERATION

1. Problem Statement

The process of curve generation develops a complete mathematical definition of the curve from a set of given data elements describing certain properties of the curve. Since the data elements usually describe some, but not all properties of the shape of the curve, the results of curve generation are not unique, but they depend also on the choice of mathematical curve representation and the properties of the generation process. Thus many different techniques exist for generating curves from given data [25].

We are interested in generating free-form curves, i.e., curves of arbitrary shape, e.g. ship lines, that cannot readily be represented in simple closed analytical form, such as by conic sections.

The problem of curve generation in the general case can be stated in terms of three categories of information (Fig. 63):

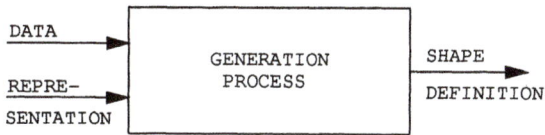

Fig. 63. Curve generation

- Data:
    A set of given data elements for the geometric shape of the curve.

- Mathematical Representation:
    A type of representation for the curve in analytic form.

- Process:
    A generation procedure to develop the curve from the data and based on the representation. The process can be described by a criterion and constraints.

Consider the following example:

"Generate a curve by interpolation of point data as a B-spline curve."

This can be restated in a more formal way:

a. Data: Discrete data points

b. Mathematical representation: B-spline

c. Process: Interpolation

The criteria and constraints needed for this problem will be discussed later.

The following section will address a systematic classification of curve generation problems based on this taxonomy. We will further demonstrate that problems stated in this simple and uniform way can all be regarded as optimization problems with constraints. Thus our viewpoint is:

"Curve generation can be formulated as optimization with constraints."

2. Classification

The following information elements serve to define curve generation problems. Any meaningful subset of these elements can be chosen to state a problem and thereby classify the task.

a. Data:

- Discrete data:
    - o Point, point set
    - o Tangent vector or vector set
    - o Curvature vector or vector set
    - o Higher derivatives
    - o Combinations, i.e., point-vector sets
- Integral shape data:
    - o Area under the curve
    - o Centroid location
    - o Higher order moments.

b. Representations:

- Types:
    - o Explicit/implicit/parametric
    - o Polynomial/non-polynomial
    - o Integer/rational polynomials
- Requirements:
    - o Continuity
    - o Parametric:$C^1, C^2, \cdots, C^n$
    - o Geometric:$G^1, G^2, \cdots,$
    - o Shape Preservation
        Positivity/monotonicity/convexity
    - o Polynomial precision

  o Controllability
     Local/Global.

- Example: Integer Parametric Polynomials

  Different polynomial bases:

  o Monomial, natural polynomial basis

$$\mathbf{r}(t) = \sum_{i=0}^{n} \mathbf{a}_i \cdot t^i \qquad (194)$$

  o Lagrange basis

$$\mathbf{r}(t) = \sum_{i=0}^{n} \mathbf{P}_i \cdot L_i(t) \qquad (195)$$

  o Hermite basis

$$\mathbf{r}(t) = \sum_{i=0}^{n} \mathbf{P}_i \cdot G_i(t) + \sum_{i=0}^{n} \mathbf{Q}_i \cdot H_i(t) \qquad (196)$$

  o Cubic spline basis

$$\mathbf{r}(t) = \sum_{i=0}^{3} \mathbf{C}_i \cdot (t - t_i)^i \qquad (197)$$

  o Bézier-Bernstein basis

$$\mathbf{r}(t) = \sum_{i=0}^{n} \mathbf{V}_i \cdot B_{i,n}(t) \qquad (198)$$

  o B-spline-Schoenberg basis

$$\mathbf{r}(t) = \sum_{i=0}^{n} \mathbf{V}_i \cdot N_{i,k}(t) \qquad (199)$$

- Rational Parametric Polynomials:

  o NURBS basis

$$\mathbf{r}(t) = \sum_{i=0}^{n} \mathbf{V}_i \cdot R_{i,k}(t). \qquad (200)$$

For identical polynomial degree, segmentation and parametrization all of these different basis representations are convertible into each other by equating coefficients, and hence are geometrically equivalent.

c. Processes:

A process for curve generation is defined by a criterion functional and by constraints. Below we give some typical examples of the criterion and constraints used in curve generation:

- Criterion:
  With the notation for the mth differential operator

$$D^m = d^m/dt^m \tag{201}$$

  the mth order 'fairness criterion' is defined by

$$L_m = \int_0^1 (D^m r(t))^2 \, dt, \tag{202}$$

  which is to be minimized.
  A well-known special case is the second order criterion which corresponds to the strain energy of flexure in a beam:

$$L_2 = \int_0^1 \left(D^2 r(t) \right)^2 dt. \tag{203}$$

- Constraints:
  o Distance constraints:
    The Euclidean distance between given data points $\mathbf{P}_i$ and the resulting curve $r(t)$, taken at the associated parameter knot t_i, weighted by w_i, and squared, must be no greater than an error tolerance ϵ. In normalized form for $n+1$ data points:

$$A = \sum_{i=0}^{n} \{w_i \left(r(t_i) - \mathbf{P}_i \right)\}^2 - \epsilon \leq 0. \tag{204}$$

  o End constraints:
    For the first and last points of the curve, tangent vectors $\mathbf{Q}_i$ or curvature vectors $\mathbf{K}_i$, $i = 0$, or n, may be given:
    Type I end conditions:

$$E = D^1 r(t_i) - \mathbf{Q}_i = 0 \tag{205}$$

    Type II end conditions:

$$E = D^2 r(t_i) - \mathbf{K}_i = 0. \tag{206}$$

  o Area constraint:
    The actual area under a curve, S, shall match a given area value, S_0:

$$F = S - S_0 = 0 \tag{207}$$

    where, for instance, for a planar curve in explicit form:

$$S = \int_a^b f(x)dx. \tag{208}$$

o Other constraints:
Many other types of constraints can be imposed in equality or inequality form.

The elements of problem description just defined constitute a very general and flexible set of ingredients for the problem classification.

3. Optimization Format

Based on the elements described in Section 2. the problem of curve generation may now be restated as an optimization problem:
"Minimize the functional L_m subject to constraints:
$$A \leq 0, \quad E = 0, \quad F = 0 \text{ etc.}"$$
The free variables in this problem are contained in the free coefficients of the curve representation of Eqs. 194-200.

This constrained optimization problem can be reformulated as a free variational problem based on the Euler-Lagrange approach. If we introduce a function

$$I = L_m + \lambda A' + \mu E + \nu F \tag{209}$$

where λ, μ, ν are known as Lagrangian multipliers, and $A' = A + d^2$, d^2 is a slack variable, then the problem is to minimize the unconstrained functional I. In this form the additional unknowns are the Lagrangian multipliers and the slack variables, which are introduced to transform inequality into equality constraints.

This variational formulation can sometimes be solved analytically (via Euler's equation of variational calculus) or at least numerically.

The development has demonstrated how curve generation can be approached from the viewpoint of an optimization problem or also by means of variational calculus. Several examples will follow to illustrate this approach.

3.1. Buczkowski's Approach

L. Buczkowski [12] at TU Gdansk was apparently the first to apply a variational approach to the generation of ship curves. As an example of his general method let us consider the generation of a planar curve (e.g. a waterline), represented by an explicit polynomial $y(x)$.

In this example Buczkowski proceeds to generate a 'fair' waterline curve $y(x)$ from a fairness criterion, a constraint on the waterplane area A, and given end conditions (offsets).

The fairness criterion is based on the strain energy of flexure in the curve (beam), which is linearly approximated as:

$$U = \int_a^b y''(x)^2 \cdot w(x)dx = \min \tag{210}$$

where $w(x)$ is a weighting function, for instance, the beam stiffness $EI(x)$.

The area constraint, Eq. 208, is

$$S = \int_a^b y(x)dx = S_0 \quad \text{(given area)} \tag{211}$$

or

$$F = S - S_0 = 0.$$

Using the Euler-Lagrange approach, the problem:
                    "Minimize U subject to $F = 0$"
has the free variational form
                    "Minimize $I = U + \nu F$."
In variational calculus the extreme value problem

$$I = \int_a^b f(x, y, y', y'')dx = \min \tag{212}$$

has the equivalent Euler equation

$$\frac{\partial f}{\partial y} - \frac{d}{dx}\frac{\partial f}{\partial y'} + \frac{d^2}{dx^2}\frac{\partial f}{\partial y''} = 0. \tag{213}$$

Therefore, in our example,

$$I = \int_a^b \left(y''(x)^2 \cdot w(x) + \nu y(x) \right) dx - \nu S_0 \tag{214}$$

has the Euler equation

$$\nu + \frac{d^2}{dx^2}\left(2y''(x)w(x)\right) = 0. \tag{215}$$

We solve for $y(x)$ by quadrature:

$$y(x) = \int_a^x \int_a^{\bar{x}} \left((k_4\bar{x}^2 + k_3\bar{x} + k_2)/w(\bar{x})\right) d\bar{x}dx + k_1 x + k_0 \tag{216}$$

where the integration constants k_0, k_1 are determined from curve end conditions. In the special case of constant stiffness function $(w(x) = \text{const})$ the double integral would yield a polynomial of fourth degree. This polynomial is the fundamental solution for a curve of optimum fairness measure U and given area, S_0. The coefficients k_2, k_3, k_4, which contain ν, can be determined numerically from the area constraint and two further end conditions.

3.2. Spline Interpolation

The classical spline interpolation problem may be regarded as a variational problem, too. Given some point data $\mathbf{P}_i$, $i = 0, \cdots, n$, to be interpolated by a piecewise continuous parametric polynomial $\mathbf{r}(t)$, as in Fig. 64. We require again the fairness criterion L_2 to be minimized:

$$L_2 = \int_a^b \left(\frac{d^2\mathbf{r}(t)}{dt^2} \right)^2 dt = \min. \tag{217}$$

The constraints are point interpolation conditions of the form

$$\mathbf{r}(t_i) - \mathbf{P}_i = 0, \quad i = 0, \cdots, n \tag{218}$$

or for all points

$$A = \sum_{i=0}^{n} (\mathbf{r}(t_i) - \mathbf{P}_i)^2 = 0. \tag{219}$$

The Lagrangian form

$$I = L_2 + \lambda A = \int_a^b \mathbf{r}''(t)^2 dt + \lambda \sum_{i=0}^{n} (\mathbf{r}(t_i) - \mathbf{P}_i)^2 = \min \tag{220}$$

has the Euler equation

$$\frac{d^4\mathbf{r}(t)}{dt^4} = 0. \tag{221}$$

It follows by quadrature that $\mathbf{r}(t)$ is a cubic polynomial in each segment of the curve. Thus the cubic spline is the solution to an optimization problem which minimizes the L_2-norm and interpolates all data points. This solution was already derived by Holladay [30].

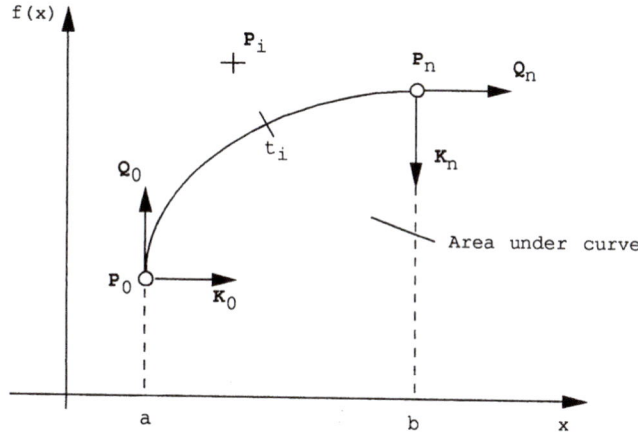

Fig. 64. Planar curve notation

The same result is obtained for a curve that approximates a data set $\mathbf{P}_i$ with tolerances ϵ_i.

3.3. Different Criteria

Meier [43] has investigated the influence of the choice of different criteria upon the shape of the resulting optimized curve. He compared the effects of replacing the L_2-fairness measure by L_3 and L_4 measures:

$$L_3 = \int_a^b \left(\frac{d^3 \mathbf{r}(t)}{dt^3} \right)^2 dt = \min \qquad (222)$$

$$L_4 = \int_a^b \left(\frac{d^4 \mathbf{r}(t)}{dt^4} \right)^2 dt = \min. \qquad (223)$$

These norms, when compared to the properties of a continuous elastic beam, are related to the curvature or strain energy (L_2), change of curvature or shear force (L_3), and 'acceleration' of curvature or continuous load (L_4).

Fig. 65 gives an example. A data set of five points is interpolated by a polynomial of degree 10. This degree is intentionally chosen well above the number of point constraints in order to have enough freedom, i.e., free coefficients, for shape optimization (underdetermined case). The results show flat spots in the L_2 solution since the curvature integral is minimized. This is frequently not desired. The L_3 solution has more pleasant, gentle changes in curvature. The L_4 result may not be attractive due to its extreme bulge between data points 0 and 1.

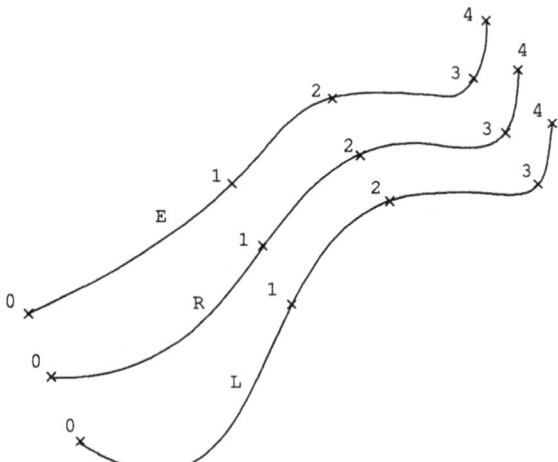

Fig. 65. Curve interpolation with different fairness measures (Degree 10, $L_2 \equiv E$: energy, bending moment, $L_3 \equiv R$: change of curvature, shear force, $L_4 \equiv L$: continuous load.)[43],[44]

From the work of Meier, we can thus conclude:

-   The choice of criterion does affect shape characteristics.

- A surplus of free coefficients enables the curve freely to respond to the criterion.

- Ship lines designers are often interested in gradual changes in curvature, also for hydrodynamic reasons. They should then be using the L_3 measure [44].

4. Fair Bézier Curves with Constraints

A curve generation problem that frequently arises in ship design, is the interpolation of some data points combined with an area constraint and end conditions. This problem is here approached for the design of a planar curve $r(t)$ represented by a Bézier polynomial of degree n. The degree is chosen to be well above the number of constraints.

The Bézier curve is given by:

$$r(t) = \sum_{i=0}^{n} \mathbf{V}_i \cdot B_{i,n}(t). \tag{224}$$

The constraints are

- Interpolation conditions $(m+1)$:

$$r(t_i) - \mathbf{P_i} = 0, i = 0, \cdots, m \tag{225}$$

- End conditions:
  The final legs of the defining polygon are parallel to the given end tangent vectors $\mathbf{Q}_0$ and $\mathbf{Q}_n$:

$$\begin{aligned} \mathbf{V}_1 - \mathbf{V}_0 &= \mu_1 \mathbf{Q}_0 \\ \mathbf{V}_n - \mathbf{V}_{n-1} &= \mu_2 \mathbf{Q}_n \end{aligned} \tag{226}$$

  where μ_1, μ_2 are unknown free scalar factors.

- Area condition:
  The area S of a sector between the origin and the curve $r(t)$ between parameters t_0 and t_1 (Fig. 66) is described by the integral of a cross product:

$$S = \frac{1}{2} \int_{t_0}^{t_1} r(t) \times \frac{dr(t)}{dt} dt \tag{227}$$

  where the symbol $\times$ denotes the cross product of two vectors $r_1 = (x_1, y_1)$ and $r_2 = (x_2, y_2)$ as follows

$$\mathbf{r}_1 \times \mathbf{r}_2 = \begin{vmatrix} x_1 & y_1 \\ x_2 & y_2 \end{vmatrix} \tag{228}$$

  This area is equated to a given area S_0. The area S can be developed into an expression of the form

$$S = \frac{n}{2} \sum_{i,j=0}^{n} d_{ij} (\mathbf{V}_i \times \mathbf{V}_j) = \frac{n}{2} \Omega \tag{229}$$

where the d_{ij} are coefficients based on integrals of Bernstein polynomial bi-products by substituting the Bézier curve. For details see [55].

The area constraint is thus:

$$F = S - S_0 = \frac{n}{2}\Omega - S_0 = 0. \tag{230}$$

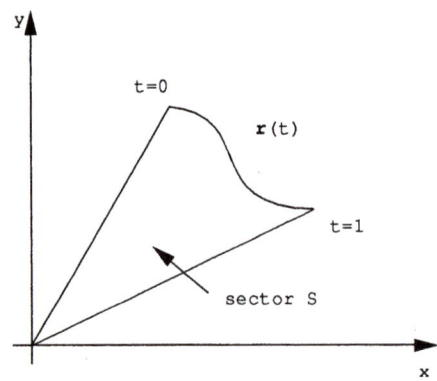

Fig. 66. Definition of sector area under curve

- Fairness constraint:
      The fairness criterion chosen here is the L_2 measure [51].

$$L_2 = \int_0^1 \left(\frac{d^2\mathbf{r}}{dt^2}\right)^2 dt = \min. \tag{231}$$

The variational problem thus amounts to:

$$I = L_2 + \mu_1 E_1 + \mu_2 E_2 + \nu F = \min \tag{232}$$

with E_1, E_2 as end constraints, if present and μ_1, μ_2, ν are Lagrangian multipliers.

The free variables in I consist of the free Bézier points $\mathbf{V}_i$ (surplus: $r = n + 1 - m - 1 = n - m$, potentially minus end conditions) as well as the Lagrangian multipliers ν etc. The minimum conditions thus are of the type:

$$\frac{\partial I}{\partial \mathbf{V}_i} = 0 \quad, \frac{\partial I}{\partial \nu} = 0 \quad \text{etc.} \tag{233}$$

This yields a nonlinear system of equations for the unknowns, which can be solved numerically. The complete derivation and solution procedure is documented in [55].

A few examples will illustrate the approach. Fig. 67 shows a quarter circle specified by four equidistant data points with an area constraint of $S_0 = 0.785$ and a polynomial degree of $n = 14$. It closely resembles the true circle. Changing the polynomial degree within reasonable limits does not alter the shape very much. The measure L_2 changes only slightly:

Table 6. L_2 measure for different degree polynomials

Degree	n	6	9	12
Measure	L_2	5.93	5.64	5.53

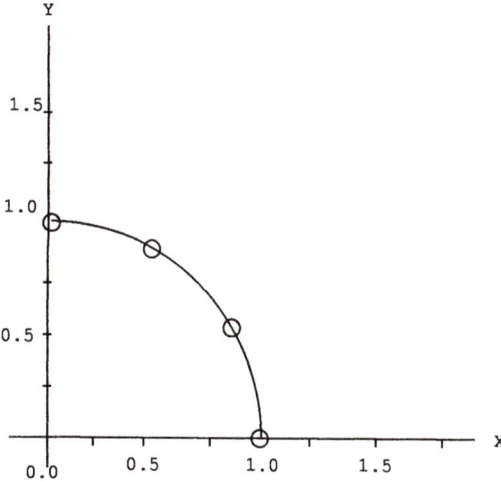

Fig. 67. Quarter circle approximated by polynomial of degree $n = 14$, interpolated through four data points, area $S_0 = 0.785$, tangent vectors $\mathbf{Q}_0 = (0, 0.1)$, $\mathbf{Q}_1 = (-0.1, 0)$ [55]

In Fig. 68 and Table 7. for the same data points and a change in area S_0 for polynomial degree $n = 10$ the criterion measure changes drastically:

Table 7. Area constraint for different degree polynomials

Area	S_0	0.785	1.047	1.309
Measure	L_2	5.55	749.0	2260.0

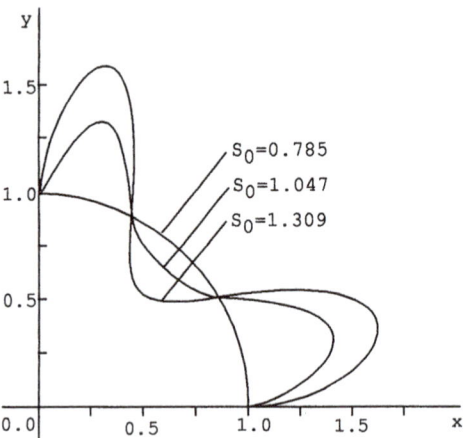

Fig. 68. Variation in area constraint: $S_0 = 0.785; 1.047; 1.309$, polynomial degree $n = 10$, same point data set as in Fig. 67, but without end vectors [55]

Fig. 69 presents the interpolated Bézier curves for the data of Fig. 65 with $S_0 = 0.25$ and degrees $n = 6$, 8 and 10. The fairness measure varies as follows:

Table 8. Degree constraint

Degree	n	6	8	10
Measure	L_2	108.2	56.7	53.7

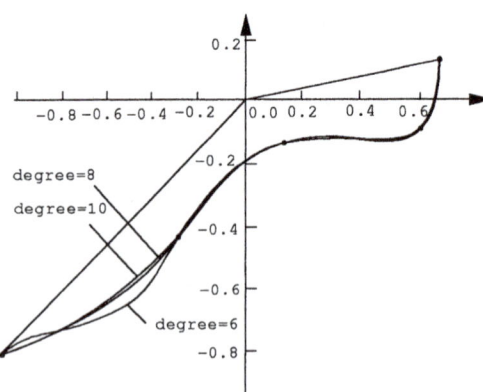

Fig. 69. Variation in polynomial degree n=6, 8 and 10 for five data points of Fig. 65 and area constraint of S_0=0.25 [55]

Fig. 70 finally shows a variation in area $S_0 = 0.25, 0.26, 0.27, 0.28$ for $n = 10$ with the results:

Table 9. Area constraint

Degree	S_0	0.25	0.26	0.27	0.28
Measure	L_2	53.7	56.6	61.3	68.0

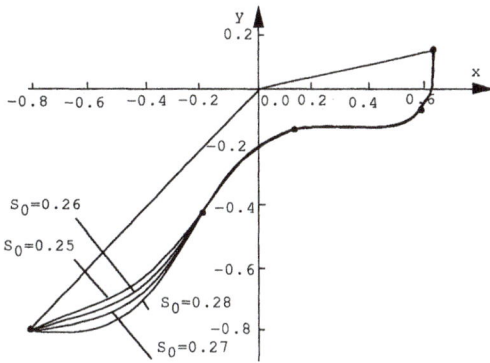

Fig. 70. Variation in area constraint: S_0=0.25; 0.26; 0.27 ; 0.28. polynomial degree n=10

5. Summary

The process of curve generation is based on geometric input data, a mathematical representation, a criterion function and constraints. This description serves to classify many different types of generation procedures. The problem can also be stated in variational form and hence as an optimization problem. For each problem type a fundamental, 'best' solution is derived. This approach leads to numerical solutions for a large class of generation problems. Examples, mainly of interpolation with constraints, illustrate the methodology.

CHAPTER 5
SHIP CURVE DESIGN

1. Problem Statement

The geometry of ship hull surfaces is traditionally designed and represented by a set of curves, called the ship lines plan. This lines plan primarily consists of planar curves, viz. waterlines, sections and buttock lines, lying in orthogonal planes parallel to the coordinate axes. In addition diagonals, i.e. planar intersections of the ship surface in inclined planes, are used for checking purposes. Furthermore a few spatial, generally non-planar curves like side of deck (sod), knuckle lines etc. are needed to delimit the ship surface and its regions. Sometimes the boundary curves of flat of side (fos) and flat of bottom (fob) are also given. All curves in the lines plan combine together to form a mesh of curves from which points on the ship's surface can be obtained by interpolation. The surface is thus defined only indirectly.

This traditional approach was developed by ship designers and loftsmen a few centuries ago who were relying on draughting and fairing tools, such as elastic splines (also called battens), by means of which planar curves of fair shape could be drawn. However, even where surface definitions are currently available on computer systems, the designer favours the design of a surface in terms of a curve mesh rather than by direct surface generation techniques. This is as the designer's knowledge about the desired shape can best be expressed by means of form parameters for individual, characteristic ship curves. Similar technically relevant data for surface shape properties are intuitively difficult to find.

Nevertheless a few modern CAD systems exist in which ship surfaces can also be directly defined. Methods for ship surface definition are further discussed in Chapter 9.

The current problem to be addressed here can thus be stated as follows:
"Develop a set of curves, most of them planar curves, such that a mesh of ship lines
    is formed by which a ship surface geometry of some desired shape is defined."
The desired shape characteristics are usually described by form parameters for the individual curves as well as by a few global shape properties of the ship hull as a whole.

The principal methods for developing such a set of ship curves are the following [52]:

- Form parameter design:
    Ab initio design of a lines plan based on form parameter design of individual curves.

- Distortion:
    Transformation of a given hull form definition by distortion operations to achieve new form data.

- Interpolation:
  Derivation of a new hull form by interpolation in a given systematic series of hull forms, such as Taylor series or Series 60.

This chapter will concentrate on the most general approach of form parameter design in Sections 2 to 4. The other two procedures will be discussed in Section 5.

2. Form Parameter Design of Basic Curves

It is feasible to develop a lines plan from a small set of basic curves, i.e. a minimal set of curves conceived from the requirement that all other ship lines can be deduced from this set and that the essential main shape characteristics of the hull can be controlled by means of this set. Such a set of basic curves can be chosen as:

- Sectional Area Curve (SAC)

- Design Waterline (DWL)

- Main Deck, top view

- Lateral Plan (side view of deck, stem and stern profile, keel)

Fig. 71 illustrates these basic curves in a Cartesian coordinate system whose origin is amidships, in the centreplane and at base line level. The significant form parameters defining each basic curve are given in these diagrams. The notation for Fig. 71 is as follows:

Table 10. Notation for ship form definition

Principal Dimensions and Coefficients	
L_{WL}	length of waterline
L_{PP}	length between perpendiculars
B	beam
T	draught
D	depth
∇	volume of displacement
$C_B = \nabla/(L_{PP}BT)$	block coefficient
A_M	midship section area
$C_M = A_M/(BT)$	midship section coefficient
$C_P = \nabla/(L_{PP}A_M)$	prismatic coefficient
$\quad = C_B/C_M$	
$C_V = \nabla/L^3$	volumetric coefficient

Table 11. Notation for sectional area curve

Sectional Area Curve	
$A_S(x)$	section area below DWL, plotted against length
L_{SAC}	length of SAC
L_E	length of SAC entrance
L_M	length of parallel midbody
L_R	length of SAC run
∇_E	volume of entrance
∇_M	volume of parallel midbody
∇_R	volume of run
X_E	long. centroid of entrance
X_M	long. centroid of midbody
X_R	long. centroid of run
X_{LCB}	long. centre of buoyancy
A_{SF}	sectional area, fore perpend. (FP)
A_{SA}	sectional area, aft perpend. (AP)
$C_{PE} = \nabla_E/(L_E A_M)$	prismatic coefficient, entrance
$C_{PR} = \nabla_R/(L_R A_M)$	prismatic coefficient, run

Table 12. Notation for design waterline

Design Waterline	
$y(x)$	half-beam of waterline
l_E	length of waterline entrance
l_M	length of parallel part of DWL
l_R	length of waterline run
A_{WP}	waterline area
A_{WPE}	waterline area, entrance
A_{WPM}	waterline area, midbody
A_{WPR}	waterline area, run
X_{LCF}	longitudinal centre of flotation
y_{WPA}	half-beam at AP
$C_{WP} = A_{WP}/(L_{WL}B)$	waterplane coefficient
$C_{WPE} = A_{WPE}/(l_E B)$	waterplane coefficient, entrance
$C_{WPR} = A_{WPR}/(l_R B)$	waterplane coefficient, run
I_T	transverse moment of inertia
$C_{IT} = I_T/(B^3 L/12)$	coeff of transverse moment of inertia

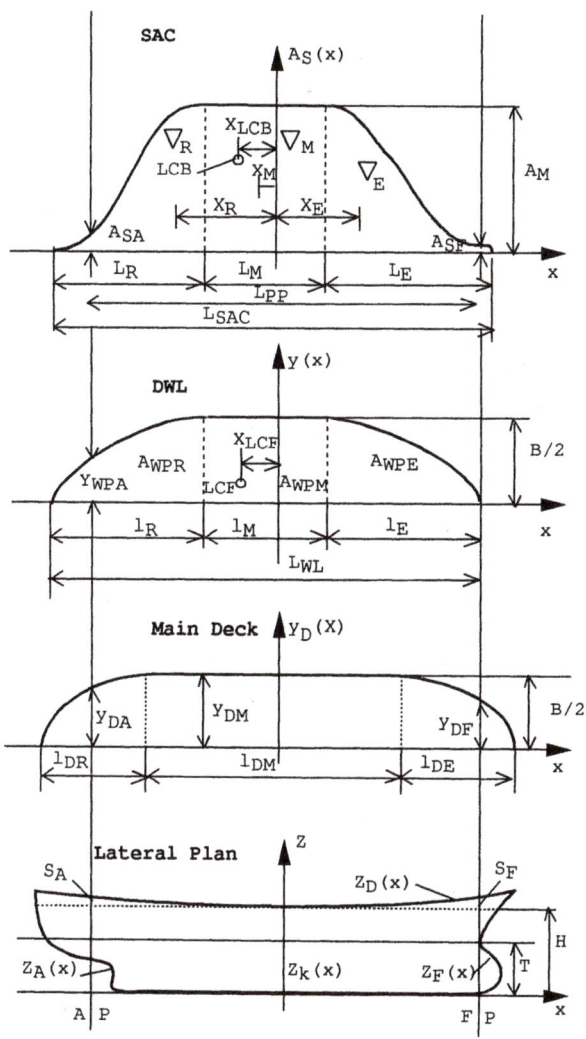

Fig. 71. Basic curves of ship form definition

Table 13. Notation for keel and end profiles

Keel and End Profiles	
$Z_k(x)$	keel contour, side view
$Z_F(x)$	stem profile, side view
$Z_A(x)$	stern profile, side view

Table 14. Notation for main deck

Main Deck	
$y_D(x)$	half-beam, main deck
l_{DE}	length of entrance, main deck
l_{DM}	length of parallel part, main deck
l_{DR}	length of run, main deck
y_{DF}	half-width, main deck, at FP
y_{DM}	half-width, main deck, amidships
y_{DA}	half-width, main deck, at AP
$Z_D(x)$	sheer curve of deck (centreplane)
S_F	sheer at FP
S_A	sheer at AP

Each of the basic curves can be designed from the set of form parameters which is given by its set of symbols, sometimes supplemented by very few other pieces of information. Many curves are designed in segments (e.g. entrance, parallel midbody, run). For these curves the form parameters of the segments are connected with the global properties of the curve by requirements on principal characteristics and coefficients. For example, for the Sectional Area Curve:

$$\nabla = \nabla_E + \nabla_M + \nabla_R \tag{234}$$

$$X_{LCB} = (\nabla_E x_E + \nabla_M x_m + \nabla_R x_R)/\nabla \tag{235}$$

$$L_{SAC} = L_E + L_M + L_R \tag{236}$$

and for the Design Waterline (DWL):

$$A_{WP} = A_{WPE} + A_{WPM} + A_{WPR} \tag{237}$$

$$X_{LCF} = (A_{WPE}x_{WPE} + A_{WPM}x_{WPM} + A_{WPR}x_{WPR})/A_{WP} \tag{238}$$

$$L_{WL} = l_E + l_M + l_R \tag{239}$$

where x_{WPE} is the longitudinal centre of entrance, etc. These relationships act as constraints on the shape of these curves.

Generalizing from these examples it is possible to derive a standardized curve design format for planar ship curves, as shown in Fig. 72. According to this format, all design problems are special cases of the following task:

"Develop a planar curve of one or several segments (usually no more than three) such that the curve meets its global form requirements and each segment complies with its regional and local form parameters."

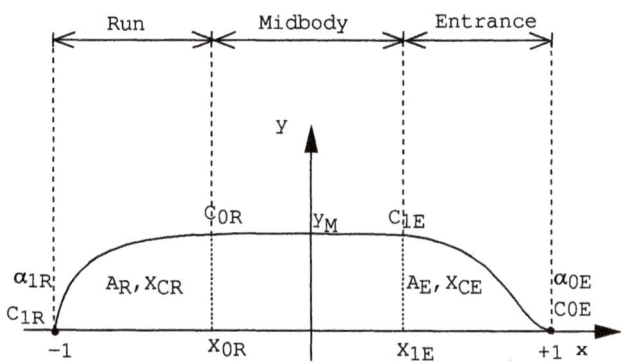

Fig. 72. General format of ship curve definition (with three segments)

The following form parameters are relevant (see Fig. 72):

- Global form parameters:

  o Principal dimensions

  o Area of curve (A_{WP} etc.)

  o Centroid of curve (x_{LCF} etc.)

- Segment form parameters:

  o Segment length (given by x_{0R}, x_{0E} etc.)

  o Segment width (given by y_M)

  o Segment area (A_R, A_E etc.)

  o Segment centroid (x_{CR}, x_{CE} etc.)

- Local form parameters:

  o Defined at segment end points:

  o Slopes, tangent vectors (α_{0E}, α_{1R} etc.)

  o Curvatures ($C_{0E}, C_{1E}, C_{0R}, C_{1R}$)

In practice, all or any subset of these form parameters may be given. The essential core of the problem which repeats itself many times is the definition of a curve segment with given area, centroid and end constraints. This problem is of the type for which a flexible solution method was presented in Chapter 4, Section 4.

3. Design of Body Plan

When the basic curves of Section 2 are given, the sections in the body plan can be readily derived. Fig. 73 illustrates the design of a section at some particular station x. For each station the following information can be deduced from the basic curves:

- Section area under DWL $(A_S(x)/2)$

- Keel (or stem/stern profile) point

- Waterline beam

- Deck half-width $(y_D(x))$

The deck edge may be given indirectly via the sheer curve and a camber curve. Wherever the camber curve intersects with the given deck half-width, a point at the edge of the side of deck with the hull section is found.

For stations near the midship section the following additional information may be relevant and given (Table 15 and Fig. 73):

Table 15. Notation for section in body plan

a_{FK}	half-width of flat keel
α	deadrise angle
r_b	bilge radius

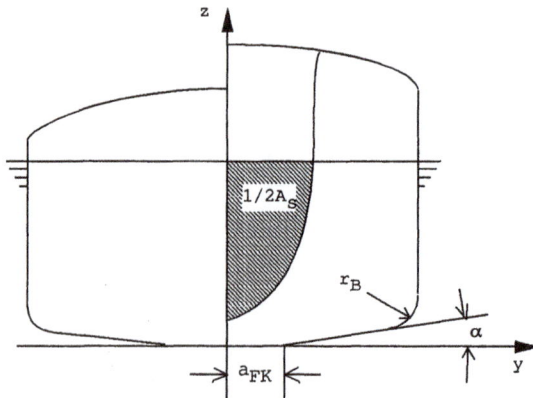

Fig. 73. Definition of section in body plan

In special cases the station is further subdivided by knuckle lines (as in chine hulls) or curvature discontinuities. In these situations the section design procedure must be modified to include several segments.

In the normal case only two segments are sufficient, the underwater part below the DWL where the section curve is fitted through the end points with an area constraint, and the abovewater part between the DWL and main deck where the end points are interpolated. In both cases it is usually of advantage if end tangent vectors (or directions) can also be given. Obviously the tangent direction should normally be continuous at the DWL, preferably even the curvature at this segment transition, too. If this kind of continuity is to be ensured, then the basic curves should be supplemented by further 'angle curve' or 'tangent vector curve' information.

To summarize, the design of ship sections in the body plan can be treated as the generation of a curve of two (or more) segments. The segments are subject to end conditions and, in the underwater part, also to area constraints. The problems which arise in this development are again of the type 'curve generation with constraints' as described in Chapter 4, Section 4.

Example

Fig. 74 shows an example of a ship underwater section whose shape was generated by the method of Chapter 4.4. The section was generated as a curve represented by a Bézier polynomial of degree 19 [39]. The constraints are:

- Interpolation conditions:
    Interpolation of eight given offset data points

$$\mathbf{r}(t_k) - \mathbf{P}_k = 0, \quad k = 1, \cdots, 8.$$

- End conditions:
    Curve parallel to end tangent vectors

$$\mathbf{V}_1 - \mathbf{V}_0 = \mu_1 \mathbf{Q}_0; \quad \mathbf{Q}_0 = (1.0, 0.0)$$

$$\mathbf{V}_n - \mathbf{V}_{n-1} = \mu_2 \mathbf{Q}_n; \quad \mathbf{Q}_n = (0.302, 1.620)$$

- Area constraint:
    Sectional area fixed

$$S - S_0 = 0; \quad S_0 = 23.915.$$

- Fairness criterion:
    Minimization of L_2 measure

$$L_2 = \int_0^1 \left(\frac{d^2\mathbf{r}}{dt^2} \right)^2 dt = \min.$$

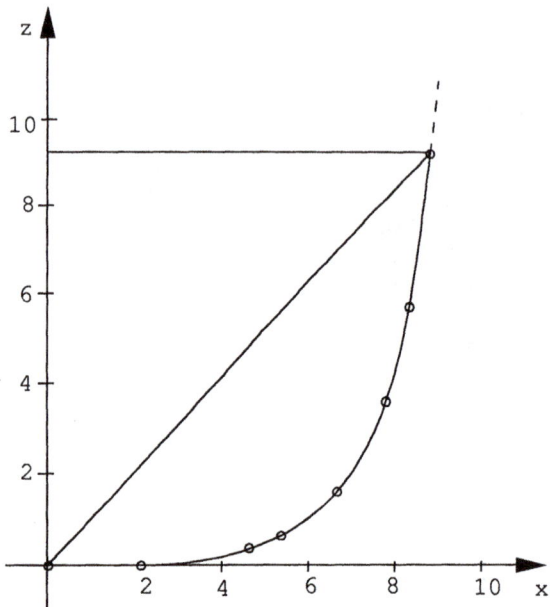

Fig. 74. Ship section interpolated through 8 given offset points [55]

4. Lines Plan Development

The development of the complete lines plan can follow as soon as the basic curves and body plan are defined to the designer's satisfaction. Even if all of the initially given form requirements are met, there are sometimes flaws in the resulting shapes because the form parameters may not be appropriately attuned to start with. It takes judgement and often some iteration in order to harmonize all requirements [67].

The iterative nature of this process is shown in Fig. 75. According to this flow chart the process proceeds in several steps and iterations from the design of the basic curves and body plan sections (as in the previous example) to the examination of numerous derived checking curves. The checking curves may be further waterlines, buttocks and diagonals. All of these curves can be derived by interpolation from the existing curve sets. They play the role of derived information in this context. If any results are to be modified, then the basic curves or body plan will be changed first before the checking curves are rederived. This sequence is chosen in order to keep the form parameters consistently under control.

In manual design and in some CAD systems one can, of course, modify any curve in the ship lines curve mesh and update the other curves accordingly. This may be required if one, or more, of the sections do not appear fair. However, in the form parameter design of an initial hull shape this may not be opportune because it would jeopardize the form parameters.

When all primary and derived curves of the lines plan are accepted by the designer, the lines plan is complete. It then provides a hull form definition by a curve mesh. This mesh can be interpolated by a surface as will be discussed in Chapters 8 and 9.

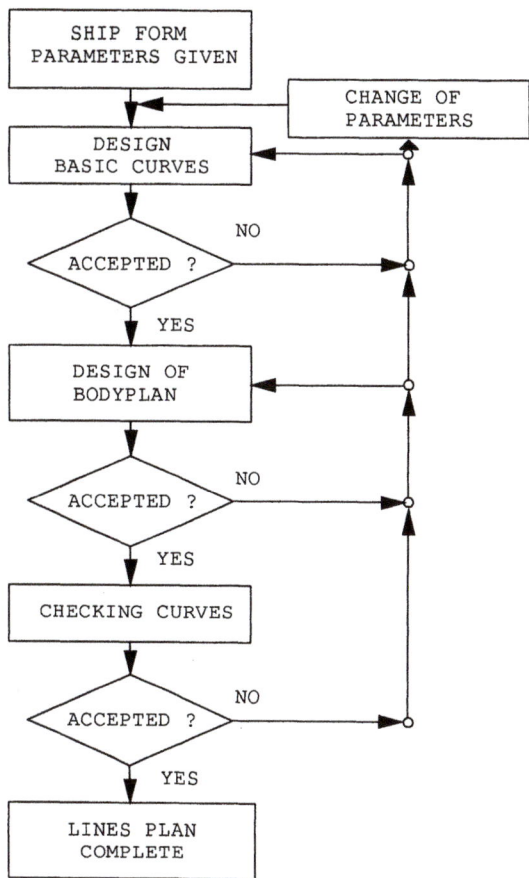

Fig. 75. Flow chart for lines plan development by form parameter method

5. Lines Distortion Approach

Hull form definitions can also be derived from an existing hull by distortion transformations or from a hull form series by interpolation.

This distortion approach transforms a given ship curve, usually the sectional area curve, by shifting its ordinates along the abscissa axis in order to achieve a new area and/or centroid location. As a simple example we will consider a linear distortion of the sectional area curve (Lackenby [40]).

Let us look at a given sectional area curve $A_S(X)$, Fig. 76. For the forward half of the SAC shown in the figure the notation is defined as follows:

Table 16. Sectional area curve notation

$A_S(X)/A_M$	Dimensionless section area
X	Dimensionless abscissa
L_P	Parallel length of half-body
$\overline{X}$	Centroid of SAC of half-body
ΔX	Longitudinal shift of abscissa
ΔL_P	Lengthening of parallel part
H	Centroid of added SAC area of half-body

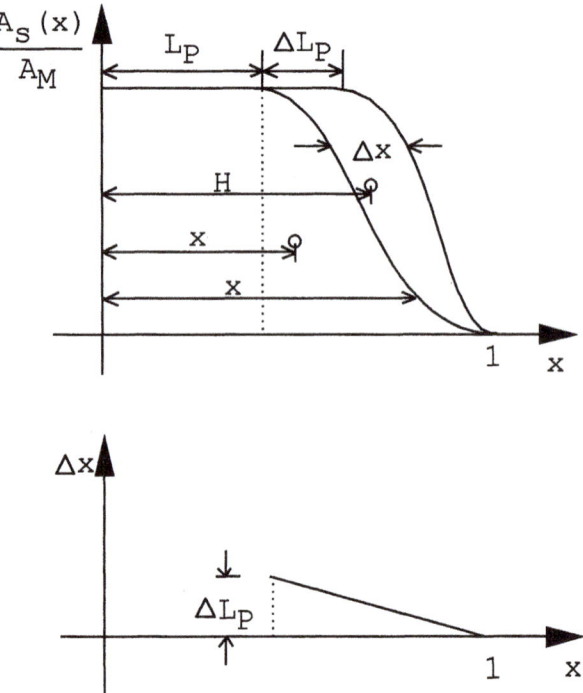

Fig. 76. Sectional area curve with distortion parameters

The distortion operation is linear, i.e., it is assumed that the longitudinal abscissa shift ΔX is a linear function of X whose ordinate at $X = L_P$ is equal to ΔL_P, Fig. 76, lower part. In this case the distortion of the half-body is governed by only one free parameter ΔL_P. This freedom can be used to change the area of the SAC of the half-body or, which is equivalent, to raise the prismatic coefficient C_{PF} by ΔC_{PF}. With reference to Fig. 77, our notation is:

Table 17. Distortion notation

C_{PF}	area under SAC, forebody before distortion
ΔC_{PF}	change of C_{PF} by distortion, hence change of area under SAC, forebody

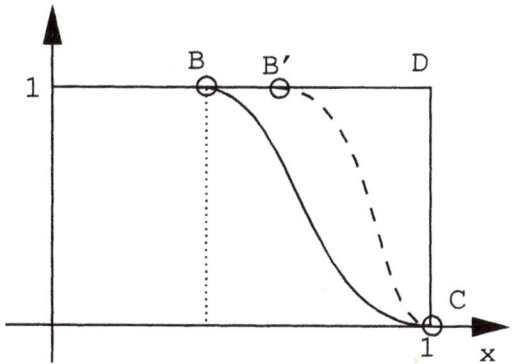

Fig. 77. Derivation of shift function ΔX

According to Fig. 77 we are making a linear shift, which moves the shoulder of the parallel part from B to B' and which we propose to be:

$$\Delta X = \frac{\Delta C_{PF}}{1 - C_{PF}}(1 - X) \tag{240}$$

and, in particular:

$$\Delta L_P = \frac{\Delta C_{PF}}{1 - C_{PF}}(1 - L_P). \tag{241}$$

This yields exactly the desired change ΔC_{PF}. The proof is based on Fig. 77:

The ratio of these areas must be equal to the ratio of the ordinates of these areas before and after the distortion, as measured from the base line of CD. Hence

$$\frac{B'CD}{BCD} = \frac{1 - (C_{PF} + \Delta C_{PF})}{1 - C_{PF}} = 1 - \frac{\Delta C_{PF}}{1 - C_{PF}} = \frac{1 - (x + \Delta x)}{1 - x} = 1 - \frac{\Delta x}{1 - x} \tag{242}$$

and consequently,

$$\Delta X = \frac{\Delta C_{PF}}{1 - C_{PF}}(1 - X), \quad \text{q.e.d.} \tag{243}$$

The same transformation can be applied to the aft half-body. Combining the two area changes it is possible to correct a given SAC in two free parameters, e.g., its prismatic coefficient and longitudinal centroid. If the given SAC is to be changed by ΔC_P and ΔX_{LCB}, it can be shown that two halves must be corrected in the following way:

$$\Delta C_{PF(A)} = \frac{2\Delta C_P \left(H_{A(F)} \pm X_{LCB}\right)}{H_F + H_A} \pm \frac{2\Delta X_{LCB}(C_P + \Delta C_P)}{H_F + H_A} \tag{244}$$

where

$$H_{F(A)} = \frac{C_{PF(A)}(1 - 2\overline{X}_{F(A)})}{1 - C_{PF(A)}} + \frac{\Delta C_{PF(A)}}{2(1 - C_{PF(A)})^2} \left(1 - 2C_{PF(A)}(1 - \overline{X}_{F(A)})\right) \tag{245}$$

The quantities in parentheses, (A) etc., refer to the aft half-body. This method is also known as the $(1 - C_P)$ method (Lackenby [40]).

The approach has been generalized to allow quadratic and even higher order distortion operations ([40], [66]). These transformations permit a greater number of free variables of the SAC to be modified independently. Regardless of whether the distortions are linear or of higher order, the distortion must be applied with caution and limited in extent in order to avoid degeneracies of the curve.

6. Standard Series Approach

The idea of the standard series approach is simply to interpolate a desired new hull form within the variety of designs available of systematic hull form series. Hull designs developed in this way for resistance or propulsion test series offer the advantage that they are based on the principle of varying one significant hull form parameter at a time. This facilitates interpolation for new parameter combinations within the scope of the series.

Some series are derived from a single parent, others from several parent designs. Variations from the parents are usually developed by systematic distortion methods like those described in Section 5. Certain parent designs and deduced variations are documented with the published series in terms of lines plans, sectional area curves, cross curves of offsets and/or offset tables. It is straightforward, in principle, to interpolate within this body of information; in some series there are actual mathematical principles underlying the variation in hull form, such as for Taylor's sectional area curves and design waterlines.

The range of variation in the series is limited so that the forms that can be deduced from the series are subject to corresponding limitations. Unfortunately, in each series, there are only a few form parameters that can be varied independently with several form parameters being handled as dependent variables. Interpolation is then limited to the independent variables with the designer having to accept the outcome of the dependent variables. For example, Taylor's standard series [52] gives the following ship lines information:

Independent variables: $C_P, B/T, C_V$

Dependent variables: L/B

Fixed: C_M, X_{LCB}

No. of parents: One

Documentation: Sectional area curves, offset cross curves, stem and stern profiles.

It is evident that the standard series approach encompasses only some of the simpler variations in hull form with respect to the proportions of the main dimensions, fullness, and occasionally centroid location. Other limitations are that many series confine themselves essentially to the underwater part of the hull and do not define the shape or location of the main deck. Another important disadvantage is that parents are not always sufficiently similar or the documented variations are not closely enough spaced to be able to rely on the fairness of the interpolated lines. This then requires hand fairing which adds additional effort.

The main advantage, lies in the simplicity of the standard series approach. In many feasibility studies of the early design stage it is sufficient to have a rough lines design that approximates the principal form characteristics desired. This design may be regarded as a 'dummy' for several design calculations until a more elaborate lines plan is produced. Additionally, if one of the more successful variants is departed from, there is every chance that the resulting design will have favourable hydrodynamic properties. Finally, many of the lines distortion techniques may be applied to parent designs from systematic series.

7. Summary

This section describes a methodology for ship curve design, based on the form parameter design of individual curves and proceeding in a stepwise and iterative manner to a complete ship lines plan, which is made up of a mesh of curves lying on the hull surface. The basic steps in this procedure correspond to the curve generation methods discussed in Chapter 4.

Chapter 5 also reviews Lackenby's method of hull form distortion, and discusses interpolation techniques via the standard series approach.

CHAPTER 6

ELEMENTARY MATHEMATICAL PROPERTIES

OF SURFACES

1. Introduction

In Chapter 5 it was shown how a ship surface can be derived in terms of a curve mesh (the ship lines plan, incorporating the ship section curves, waterlines etc.). However, it is also beneficial to be able to model the ship with a surface mesh so that we can quantify its properties. From knowing about the mathematical properties of the surface we can then think about altering the ship surface design. Therefore, as a start, we need to define the mathematical properties of a surface.

1.1. Elementary Surface Representation

Examples of mathematical functions which represent surfaces include equations of the form

$$f(x, y, z) = 0. \tag{246}$$

This is an implicit surface representation - for instance if the equation is linear then the plane $ax + by + cz = d$ is obtained; if it is of second order then the surface will be quadric (the surface of a sphere given by $x^2 + y^2 + z^2 - r^2 = 0$). Many surfaces can be represented by implicit functions. However, there are limitations in the ability of implicit equations to represent complex surfaces: if we wished to produce the surface of a ship hull, an appropriate implicit equation would be immeasurably difficult to find. This is where parametric surface representation is of prime importance, as all surfaces can be generated from this technique; furthermore, a parametric representation of a surface facilitates the mathematical analysis of its properties.

2. Parametric Surface Representation

As with curves we can define surfaces parametrically. If we take the real parameters u and v, then the surface will be defined by the vector-valued function $\mathbf{r} = \mathbf{r}(u, v)$ where $\mathbf{r}(u, v) = (r_1(u, v), r_2(u, v), r_3(u, v))$. Thus a surface is given with its Cartesian coordinates as functions of two parameters u and v, which are defined over a certain closed interval: $u_1 \leq u \leq u_2, v_1 \leq v \leq v_2$.

We can see from above that the surface may be regarded as a mapping from a rectangular domain in the (u, v) parameter plane R^2 to a surface patch in E^3. This can be considered as a surface element from which a whole host of more complex surfaces can be built up.

An example of a parametric surface is given by the Ferguson cubic surface patch [47], which is defined by:

$$r(u,v) = \sum_{i=0}^{3}\sum_{j=0}^{3} \mathbf{a}_{ij}u^i v^j \tag{247}$$

where $\mathbf{a}_{ij}$ represents the control points of the surface and $0 \le u \le 1$, $0 \le v \le 1$.

2.1. Curvilinear Coordinates on a Surface

The parameters u and v can be regarded as defining a curvilinear coordinate system on the surface $r(u,v)$. If we fix the value of one of the variables, $u = u_0$ say, then the function $r(u_0, v)$ is now only a function of a single scalar parameter. This results in a curve which lies on the surface. By repeating this process for other values of u, and then over particular values in the v interval, we form a net of two one-parameter families of curves on the surface, which have been mapped from a regular grid in R^2 which is parallel to the coordinate axes in the (u, v) plane, as can be seen in Fig. 78. These curves over the surface are referred to as isoparametric lines (since one of the variables is constant along them).

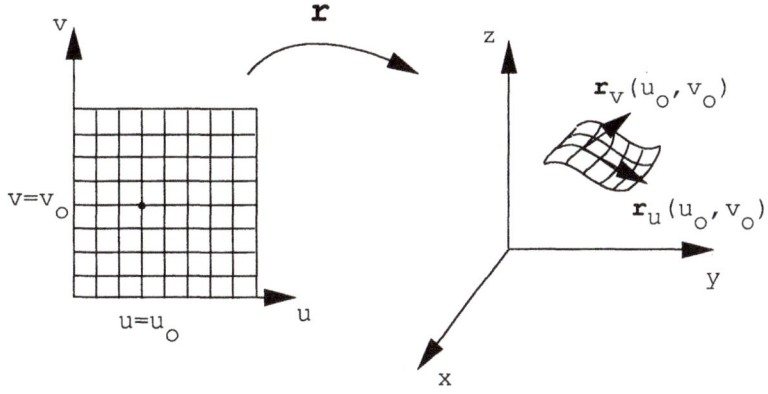

Fig. 78. The curvilinear coordinates on a surface

2.2. Coordinate Vectors for a Surface

We now define the vectors

$$\mathbf{r}_u = \frac{\partial \mathbf{r}}{\partial u}, \qquad \mathbf{r}_v = \frac{\partial \mathbf{r}}{\partial v} \tag{248}$$

to be the coordinate vectors for the surface. These are assumed to be linearly independent, i.e. they are non parallel, which implies that

$$\mathbf{r}_u \times \mathbf{r}_v \ne 0. \tag{249}$$

However the parametrization which we use is to a considerable degree arbitrary. We can easily replace the parameters u and v by a new set of parameters $\bar{u}$ and $\bar{v}$ given by $\bar{u} = \bar{u}(u,v), \bar{v} = \bar{v}(u,v)$ provided that the Jacobian

$$\frac{\partial(u,v)}{\partial(\bar{u},\bar{v})} = \begin{vmatrix} \frac{\partial u}{\partial \bar{u}} & \frac{\partial u}{\partial \bar{v}} \\ \frac{\partial v}{\partial \bar{u}} & \frac{\partial v}{\partial \bar{v}} \end{vmatrix} \neq 0 \tag{250}$$

i.e. that

$$\mathbf{r}_{\bar{u}} \times \mathbf{r}_{\bar{v}} \neq 0. \tag{251}$$

Therefore if we wish to alter from one parametrization to a new one, we are well within our rights to do so provided that the new system also has a linearly independent set of coordinate vectors.

2.3. The Tangent Plane and Normal Vector

If $\mathbf{r}(u,v)$ is a function of u and v where u and v are in turn parametrized by t, then $\mathbf{r}(t)$ is a curve on the surface with velocity vector

$$\mathbf{r}'(t) = \mathbf{r}_u u' + \mathbf{r}_v v'. \tag{252}$$

Now, at a point $\mathbf{p}$ on the line the vector $\mathbf{r}'(t)$ is the linear sum of the vectors $\mathbf{r}_u$ and $\mathbf{r}_v$ and so lies in the plane determined by these vectors. This plane is known as the tangent plane. Furthermore, any curve which passes through the point $\mathbf{r}(t)$ has its tangent lying in the tangent plane.

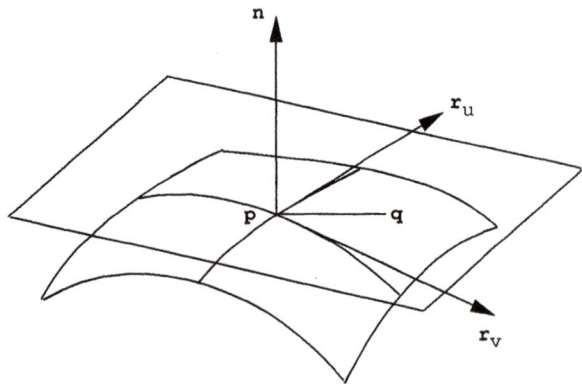

Fig. 79. The tangent plane

Therefore, if we consider a general vector on the tangent plane as $\mathbf{q} - \mathbf{p}$ where $\mathbf{q}$ is a point lying in the plane as in Fig. 79, then the equation for the tangent plane can be expressed as

$$(\mathbf{q} - \mathbf{p}) \cdot (\mathbf{r}_u \times \mathbf{r}_v) = 0. \tag{253}$$

The unit normal at any point on the parametric surface can easily be obtained. This normal is given by

$$\mathbf{n} = \frac{\mathbf{r}_u \times \mathbf{r}_v}{|\mathbf{r}_u \times \mathbf{r}_v|}. \tag{254}$$

Furthermore, if $\mathbf{r}_u$ and $\mathbf{r}_v$ are perpendicular to each other then $\mathbf{r}_u \cdot \mathbf{r}_v = 0$ and $\mathbf{r}_u, \mathbf{r}_v$ and $\mathbf{n}$ will form a local orthogonal set at the point.

Example

Consider the surface given by

$$\mathbf{r}(u,v) = \left(\cosh bu \cos v, \cosh bu \sin v, (s+3h)(u-1)^2 + (s+2h)(u-1)^3\right)$$

where $0 \le u \le 1, 0 \le v \le 2\pi$ and b, s and h are constants with $\cosh b = R$, $s < 0$, $h > 0$.

We wish to illustrate that this surface forms a blend between a vertical cylinder of unit radius centred on the z axis, and a flat horizontal plane. The blend is smooth in the sense that there is continuity of tangent plane across the junction between the blend and the adjacent surfaces.

Consider first the shape of the isoparametric line at $u = 0$:

$$\mathbf{r}(0,v) = (\cos v, \sin v, (s+3h) - (s+2h)) = (\cos v, \sin v, h)$$

which is the parametric equation of the circle $x^2 + y^2 = 1$ at a height given by $z = h$, and this obviously lies on the cylinder. Similarly at $u = 1$:

$$\mathbf{r}(1,v) = (\cosh b \cos v, \cosh b \sin v,) = (R\cos v, R\sin v, 0)$$

which is the parametric equation of the circle $x^2 + y^2 = R^2$ at $z = 0$ (the flat plane).

Thus, the edges of the blend have the correct properties in that the blend starts on the cylinder and ends on the plane.

Is the blend tangent plane continuous with the plane and the cylinder?

To acertain this we need to verify whether the direction of the normal to the blend surface approaches the directions of the normals to the cylinder and plane at the edges of the blend.

Now

$$\begin{aligned}
\mathbf{r}_u(u,v) &= (b\sinh bu \cos v, b\sinh bu \sin v, f(u)) \\
\mathbf{r}_v(u,v) &= (-\cosh bu \sin v, \cosh bu \cos v, 0)
\end{aligned}$$

where $f(u) = 2(s+3h)(u-1) + 3(s+2h)(u-1)^2$,

$$\Rightarrow \mathbf{r}_u \times \mathbf{r}_v = (-f(u)\cosh bu \cos v, -f(u)\cosh bu \sin v, b\sinh bu \cosh bu)$$

and at the edges of the blend

$$u = 0 \ : \ \mathbf{r}_u \times \mathbf{r}_v \ = \ (-s \cos v, -s \sin v, 0)$$
$$u = 1 \ : \ \mathbf{r}_u \times \mathbf{r}_v \ = \ (0, 0, Rb \sinh b).$$

Thus, we can see that when $u = 1$, $\mathbf{r}_u \times \mathbf{r}_v$ is parallel to the normal vector to the plane $z = 0$. The normal to the cylinder can be found easily. If the cylinder has a parametrization

$$\mathbf{r}(p, q) = (\cos p, \sin p, q)$$

then

$$\mathbf{r}_p \ = \ (-\sin p, \cos p, 0)$$
$$\mathbf{r}_q \ = \ (0, 0, 1)$$
$$\text{and } \mathbf{r}_p \times \mathbf{r}_q \ = \ (\cos p, \sin p, 0)$$

which is again parallel to $\mathbf{r}_u \times \mathbf{r}_v$.

Hence the surface illustrated in Fig. 80 represents a tangent plane continuous blend between a unit cylinder and a horizontal plane. In blending terms the parametric lines $u = 0, u = 1$ which bound the blend are termed the trimlines and are constrained by the shape of the surfaces being blended between. These ideas will be discussed more fully later on.

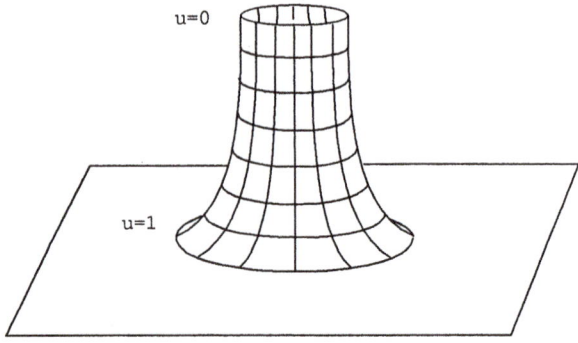

Figure 80: The surface blend between the vertical cylinder and a flat plane

3. The First Fundamental Form of a Surface

A curve in E^3 can be uniquely determined by the two local quantities curvature and torsion as functions of arc length [13]. In a similar manner a surface in E^3 can have local properties determined by quantities known as the first and second fundamental forms [73].

If we consider a regular curve on the surface $r(u,v)$ given by $r(u(t),v(t)) = r(t)$, then the vector $dr/dt = \dot{r}$ at a point P is given by

$$\frac{dr}{dt} = r_u \frac{du}{dt} + r_v \frac{dv}{dt}. \tag{255}$$

Note that $\dot{r}$ is a linear sum of r_u and r_v and hence lies in the tangent plane.

Eq. 255 can be written in a form independent of choice of curve parameter t, thus

$$dr = r_u du + r_v dv. \tag{256}$$

Now the arc length function for this curve is given by

$$
\begin{aligned}
S(t) &= \int_{t_0}^{t} \left| \frac{dr}{dw} \right| dw \\
&= \int_{t_0}^{t} \left[\left(r_u \frac{du}{dw} + r_v \frac{dv}{dw} \right) \cdot \left(r_u \frac{du}{dw} + r_v \frac{dv}{dw} \right) \right]^{1/2} dw.
\end{aligned} \tag{257}
$$

Therefore

$$S(t) = \int_{t_0}^{t} \left(r_u \cdot r_u \left(\frac{du}{dw} \right)^2 + 2 r_u \cdot r_v \left(\frac{du}{dw} \right) \left(\frac{dv}{dw} \right) + r_v \cdot r_v \left(\frac{dv}{dw} \right)^2 \right)^{1/2} dw. \tag{258}$$

The quantity

$$ds^2 = r_u \cdot r_u du^2 + 2 r_u \cdot r_v du dv + r_v \cdot r_v dv^2 \tag{259}$$

is known as the First Fundamental Form and is more commonly written as

$$I = E du^2 + 2F du dv + G dv^2 \tag{260}$$

where the first fundamental coefficients are defined by

$$
\begin{aligned}
E &= r_u \cdot r_u \\
F &= r_u \cdot r_v \\
G &= r_v \cdot r_v.
\end{aligned} \tag{261}
$$

We can interpret ds as a small increment in arc length between two points on the surface separated by du, dv. Note that since $S(t)$ is the square of a real quantity this implies that the first fundamental form is positive definitive; i.e. it is always greater or equal to zero which implies that

$$ds^2 = \frac{1}{E}(E du + F dv)^2 + \frac{EG - F^2}{E} dv^2 \geq 0 \tag{262}$$

and since $E = \mathbf{r}_u \cdot \mathbf{r}_u > 0$ it follows that

$$EG - F^2 > 0. \tag{263}$$

This can also be seen by considering the vector identity

$$
\begin{aligned}
(\mathbf{r}_u \times \mathbf{r}_v) \cdot (\mathbf{r}_u \times \mathbf{r}_v) &= (\mathbf{r}_u \cdot \mathbf{r}_u)(\mathbf{r}_v \cdot \mathbf{r}_v) - (\mathbf{r}_u \cdot \mathbf{r}_v)^2 \\
&= EG - F^2
\end{aligned} \tag{264}
$$

which implies that

$$|\mathbf{r}_u \times \mathbf{r}_v|^2 = EG - F^2 > 0. \tag{265}$$

The angle between the isoparametric lines can be found from the 1st fundamental coefficients, since

$$\mathbf{r}_u \cdot \mathbf{r}_v = |\mathbf{r}_u|\,|\mathbf{r}_v| \cos \phi \tag{266}$$

$$\Rightarrow \cos \phi = \frac{\mathbf{r}_u \cdot \mathbf{r}_v}{|\mathbf{r}_u|\,|\mathbf{r}_v|} = \frac{F}{\sqrt{EG}} \tag{267}$$

and for completeness

$$\sin \phi = \sqrt{\frac{EG - F^2}{EG}}. \tag{268}$$

From the above we can see that parametric curves are orthogonal if and only if $\mathbf{r}_u \cdot \mathbf{r}_v = 0$, i.e. $F = 0$.

Note that if the (x, y, z) coordinate system is changed (for instance by a rotation) then the first fundamental coefficients must remain unchanged since they are the scalar product of two vectors. However they will change if a new surface parametrization is used, but despite this I will remain unchanged since it represents the distance along a curve which remains fixed under coordinate transformations and reparametrization.

3.1. Surface Area

The first fundamental coefficients can also be used to find the area of some region of the surface. If we consider a small area of surface as a parallelogram defined by the u and v isoparametric lines, as in Fig. 81, then the area of the parallelogram is given by

$$dA = |\mathbf{ab}|\,|\mathbf{ad}| \sin \phi = |\mathbf{ab} \times \mathbf{ad}|. \tag{269}$$

However,

$$
\begin{aligned}
\mathbf{ab} &\approx \mathbf{r}_v dv \tag{270}\\
\mathbf{ad} &\approx \mathbf{r}_u du. \tag{271}
\end{aligned}
$$

Thus we obtain

$$dA = |\mathbf{r}_u \times \mathbf{r}_v| \, dudv = \sqrt{EG - F^2} dudv \qquad (272)$$

and the area of a region R on the surface is evaluated by integrating with respect to u and v between the appropriate limits

$$A = \int_{u_1}^{u_2} \int_{v_1}^{v_2} \sqrt{EG - F^2} dudv \qquad (273)$$

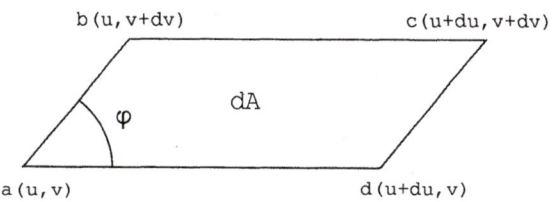

Fig. 81. Surface parallelogram

Example

We may parametrize a sphere thus:

$$\mathbf{r}(u,v) = (R \cos u \sin v, R \sin u \sin v, R \cos v)$$

where $0 \le u \le \pi, \ 0 \le v \le 2\pi$.
Then

$$\begin{aligned} \mathbf{r}_u &= (-R \sin u \sin v, R \cos u \sin v, 0) \\ \mathbf{r}_v &= (R \cos u \cos v, R \sin u \cos v, -R \sin v) \end{aligned}$$

$$\begin{aligned} \Rightarrow E &= \mathbf{r}_u \cdot \mathbf{r}_u = R^2 \sin^2 v \\ G &= \mathbf{r}_v \cdot \mathbf{r}_v = R^2 \\ F &= \mathbf{r}_u \cdot \mathbf{r}_v = 0 \end{aligned}$$

(note isoparametric lines are orthogonal since $F = 0$).
    Hence the entire surface area of a sphere equals

$$A = \int_{u=0}^{u=2\pi} \int_{v=0}^{v=\pi} \sqrt{R^4 \sin^2 v} \ dudv = 2\pi R^2 [-\cos v]_0^\pi = 4\pi R^2.$$

4. The Second Fundamental Form of a Surface

If we consider the surface in the vicinity of the point $\mathbf{r}(u_0, v_0)$ and expand this vector-valued function about the point as a Taylor series we obtain:

$$\mathbf{r}(u_0 + \Delta u, v_0 + \Delta v) = \mathbf{r}(u_0, v_0) + \mathbf{r}_u \Delta u + \mathbf{r}_v \Delta v + \frac{1}{2}(\mathbf{r}_{uu}\Delta u^2 + 2\mathbf{r}_{uv}\Delta u\Delta v + \mathbf{r}_{vv}\Delta v^2) + \cdots \tag{274}$$

Now considering the scalar function $\rho(u, v)$ defined thus:

$$\rho(u, v) = [\mathbf{r}(u, v) - \mathbf{r}(u_0, v_0)] \cdot \mathbf{n}(u_0, v_0) \tag{275}$$

where $u = u_0 + \Delta u, v = v_0 + \Delta v$ and $\mathbf{n}(u_0, v_0)$ is the unit surface normal at (u_0, v_0). The function $\rho(u, v)$ may be interpreted as the signed perpendicular distance from the tangent plane through the point $\mathbf{r}(u_0, v_0)$ to the point $\mathbf{r}(u, v)$ (Fig. 82).

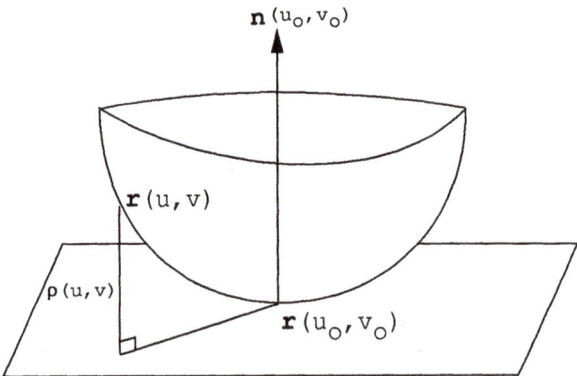

Fig. 82. Local surface in the vicinity of the point $\mathbf{r}(u_0, v_0)$

Taking the scalar product of Eq. 274 with the normal $\mathbf{n}(u_0, v_0)$ we obtain the equation for $\rho(u, v)$

$$\rho(u, v) = \frac{1}{2}\left((\mathbf{r}_{uu} \cdot \mathbf{n})\,\Delta u^2 + 2\,(\mathbf{r}_{uv} \cdot \mathbf{n})\,\Delta u\Delta v + (\mathbf{r}_{vv} \cdot \mathbf{n})\,\Delta v^2\right) + \cdots \tag{276}$$

The quantity within the brackets is termed the second fundamental formula and is often written

$$II = \frac{1}{2}\left(e\Delta u^2 + 2f\Delta u\Delta v + g\Delta v^2\right) \tag{277}$$

From the Second Fundamental Form we may determine the shape of the surface in the vicinity of the point $\mathbf{r}(u_0, v_0)$. To see this, consider the shape we consider the second fundamental form in the form

$$II = \frac{1}{e}\left((e\Delta u + f\Delta v)^2 + \left(eg - f^2\right)\Delta v^2\right) \tag{278}$$

Now the discriminant $eg - f^2$ determines whether it is possible for II (and therefore $\rho(u, v)$) to change sign in the vicinity of $\mathbf{r}(u_0, v_0)$ as Δu and Δv vary. There are four different cases to be considered:

- Elliptic case:
    A point is elliptic if $eg - f^2 > 0$. In this case the surface lies completely on one side of the tangent plane since II maintains the same sign. One example of such a surface is the sphere.

- Hyperbolic case:
    A point is hyperbolic if $eg - f^2 < 0$. By choosing suitable $\Delta u, \Delta v$ it is possible to get $\rho < 0, \rho = 0, \rho > 0$. This ensures that the surface can be on either side of the tangent plane and furthermore it passes through the plane along two distinct lines given by the solutions to the equation

$$e\left(\frac{\Delta u}{\Delta v}\right)^2 + 2f\frac{\Delta u}{\Delta v} + g = 0 \tag{279}$$

    which divide the tangent plane into four sections in which ρ is alternately positive and negative. Locally the surface is a saddle point.

- Parabolic case:
    A point is parabolic if $eg - f^2 = 0$ and $e^2 + f^2 + g^2 \neq 0$. In this case the surface is a parabolic cylinder touching the tangent plane along the line

$$e = -f\frac{\Delta v}{\Delta u} \tag{280}$$

    Planar case:
    A point is planar if $e = f = g = 0$. The surface is either a plane, or the surface is such that the method is inadequate for determining the local shape.

The different cases are illustrated in Fig. 83. Finally note that II is invariant under surface parameter and coordinate transformations.

Example

For the sphere of the previous example:

$$\mathbf{n} = (-\cos u \sin v, -\sin u \sin v, -\cos v)$$

$$\begin{aligned}
\mathbf{r}_{uu} &= (-R\cos u \sin v, -R\sin u \sin v, 0) \\
\mathbf{r}_{uv} &= (-R\sin u \cos v, R\cos u \cos v, 0) \\
\mathbf{r}_{vv} &= (-R\cos u \sin v, -R\sin u \sin v, -R\cos v).
\end{aligned}$$

Therefore

$$e = \mathbf{r}_{uu} \cdot \mathbf{n} = R\sin^2 v$$
$$f = \mathbf{r}_{uv} \cdot \mathbf{n} = 0$$
$$g = \mathbf{r}_{vv} \cdot \mathbf{n} = R$$

$$\Rightarrow \quad eg - f^2 = R^2\sin^2 v > 0$$

for all u, v. Hence for a sphere the surface is elliptic and all points lie to one side of the tangent plane (as expected).

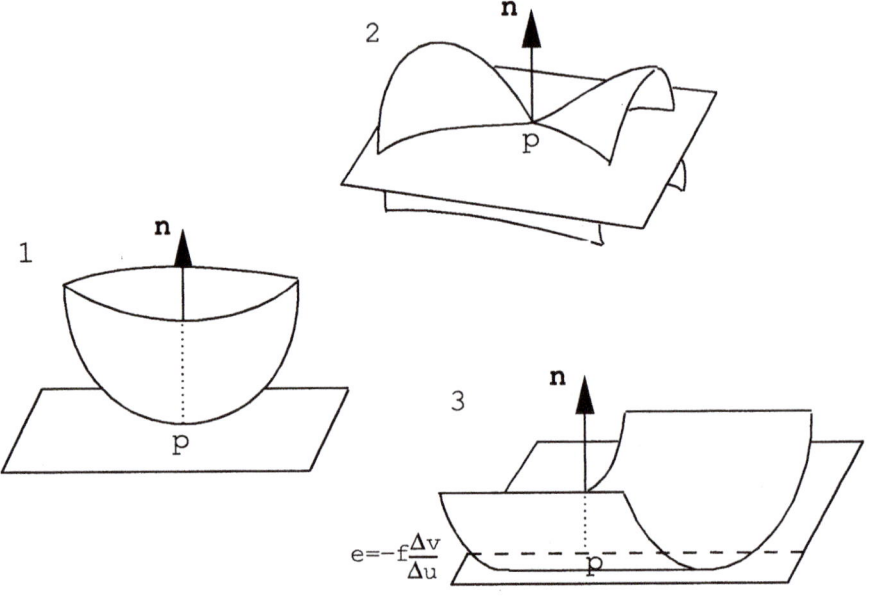

Fig. 83. The surface shapes for the cases $eg - f^2 > 0$, $eg - f^2 < 0$, $eg - f^2 = 0$

CHAPTER 7
THE PDE METHOD FOR SURFACE GENERATION

BLEND DESIGN

1. Introduction

In this chapter we will illustrate one technique used for surface generation, known as The PDE Method. The method was devised by Bloor and Wilson at the University of Leeds [6]. The method presents a different approach to surface generation from conventional Bézier, B-spline surface methods and so is included to contrast the variety of techniques open to designers. The problems the PDE method can address are not restricted to those of ship design and we shall illustrate this later in the chapter. However, in the early part of the chapter we shall deal with the details of the method, and, in particular, since the method originated from the field of blend design, how it can be used to generate blends, or fillets.

2. Blend Generation

The problem of blend generation in computer-aided design is that of being able to generate a smooth secondary surface to act as a 'bridge' between neighbouring primary surfaces [6]. The primary surfaces define the bulk of the object's shape, but the blend is perhaps required so that the object can be machined or so that sharp corners where high stresses may occur are removed. Problems in ship design may include that of producing a bridge from a structural beam to the side of the hold inside a vessel. Other notable problems include the 'blending' of a propeller blade to the hub. The problem can be viewed as a boundary-valued one in which a smooth surface is sought which satisfies a set of conditions at its boundary with the neighbouring primary surfaces.

2.1. A Surface as a Solution of a PDE

Mathematically, the calculation of the blending surface can be thought of as the solution of a partial differential equation (PDE) over some domain Ω in a parameter space, with suitably posed boundary conditions around the edges of the domain $\delta\Omega$. The combination of boundary conditions and the need for a smooth surface/solution suggests that the class of elliptic PDEs be considered to produce the solution for the surface. An obvious example of an elliptic PDE is given by Laplace's equation

$$\frac{\partial^2 \phi}{\partial x^2} + \frac{\partial^2 \phi}{\partial y^2} = \nabla^2 \phi = 0 \qquad (281)$$

where ϕ can be taken to represent the blend surface, expressed as the height $z(=\phi)$ above the (x, y) plane. By defining conditions around the boundary the surface can be determined as is illustrated in the example below:

Example

In this case x and y are independent variables with z taken as the dependent variable ϕ. Thus z gives a measure of the height of the surface and the domain Ω is such that $0 \leq x \leq 2$ and $0 \leq y \leq 2$. The boundary conditions are given by:

$$z = \left[1 - (1-y)^2\right]^{1/2} \quad \text{on } x = 0, \text{ for } 0 \leq y \leq 2$$

$$z = \left[1/4 - (y - 1/2)^2\right]^{1/2} \quad \text{on } x = 2, \text{ for } 0 \leq y \leq 1$$

$$z = \left[1/4 - (y - 3/2)^2\right]^{1/2} \quad \text{on } x = 2, \text{ for } 1 \leq y \leq 2$$

$$z = 0 \text{ on } y = 0 \text{ and } y = 2 \text{ for } 0 \leq x \leq 2. \tag{282}$$

The surface is then obtained by solving Eq. 281 with the posed boundary conditions of Eq. 282. The solution is shown in Fig. 84.

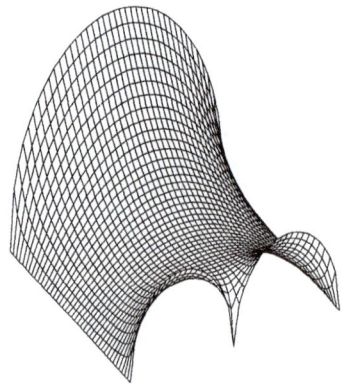

Fig. 84. The solution to Eq. 281 for the boundary conditions (Eq. 282)

While the independent variables are the Cartesian (x, y) coordinates, the surface cannot be multivalued above the (x, y) plane. Therefore, it is more appropriate to consider the surface as expressed in terms of parametric coordinates. The parameters u and v will now therefore be the independent variables in Eq. 281 and will form a parametric net of isolines in the domain over which the solution is being sought (as described in Chapter 6). This means that (x, y, z) will be functions of the parameters u and v, and hence the surface blend will be given parametrically by $\mathbf{r} = \mathbf{r}(u, v)$ where $\mathbf{r} = (x(u, v), y(u, v), z(u, v))$; in other words the solution process provides a mapping from the (u, v) parameter space into Euclidean 3-space.

3. The PDE Method

To summarize, the surface patch $r(u, v)$ is obtained by posing suitable conditions along the boundary of the domain $\delta\Omega$. From this it is natural to obtain the surface $r(u, v)$ by regarding it as the solution of a partial differential equation of the general form

$$\mathbf{D}_{u,v}^{k}(\mathbf{r}) = \mathbf{F}(u, v) \qquad (283)$$

where $\mathbf{D}_{u,v}^{k}(\ )$ is a partial differential operator of order k in the independent variables u and v. The solution of Eq. 283 gives the parametric form of the surface in the (u, v) parameter space. The choice of partial differential operator is taken to be elliptic since this ensures that the solution $r(u, v)$ can be found by posing conditions on the boundary of the domain $\delta\Omega$. The boundary conditions are posed in terms of the parameters u and v in the (u, v) parameter space and typically give a representation of the (x, y, z) coordinates.

The degree of the partial differential operator will determine the required amount of boundary data. If k is taken to be 2 then only boundary conditions representing the position of the curve need be imposed. When k is taken to be 4 then the function derivative normal to the boundaries in the (u, v) plane also can be imposed around the boundary. These first derivatives are used to control the direction of the surface normal at the edges of the patch and hence can be used to give tangent continuity between adjacent surface patches when considered in the context of blend design.

The particular equation that has been used throughout the work by Bloor and Wilson is

$$\left(\frac{\partial^2}{\partial u^2} + a^2 \frac{\partial^2}{\partial v^2} \right)^2 \mathbf{r} = 0. \qquad (284)$$

This is a biharmonic operator modified by the inclusion of the term a which has been designated the smoothing parameter [7]. This operator has many applications within the field of continuum mechanics. One of its applications is to thin plate theory where the unknown function represents the transverse displacement of a flexible thin plate bent by a load.

It should however be noted at this stage that the smoothing parameter can take a constant value or can be a function of u and v. For the case where a is constant then a simple rescaling between the directions has been achieved (this can be seen by replacing v with v/a [8]).

Therefore to obtain an appropriate surface the boundary value problem must be correctly set up and the solution sought. There are two different types of boundary value problem which we are concerned with. The first case is represented by Fig. 85a, in which positional and tangent continuity conditions must be satisfied on all boundary curves c_1, c_2, c_3, c_4. The second case is represented in Fig. 85b where the trimlines $u = u_0$ and $u = u_1$ are periodic in v, and form a closed loop of surface. In this instance boundary conditions are applied solely on c_1 and c_3. This type of surface occurs frequently in blend problems [6]; furthermore the periodicity in v is also true for many applications in free-form surface design where a single patch can be used to define a surface.

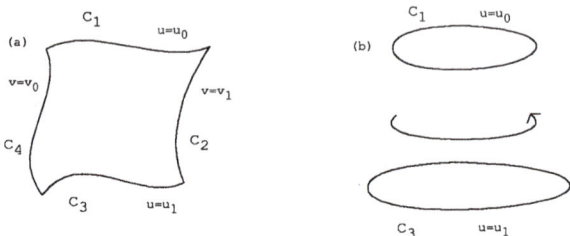

Fig. 85. The boundary value problem for a parametric surface

The partial differential operator in Eq. 284 represents a smoothing process in which the value of the function at any point on the surface is, in a certain sense, an average of the surrounding values. In this way a surface is obtained as a smooth transition between the boundary conditions imposed on both the function and its first derivative. This implies that even if sharp edges are present around the boundaries, the resulting surface obtained from Eq. 284 with the boundary conditions will be smooth. Furthermore, the solution to a fourth order PDE can also hold the property of being tangent plane continuous across the surface boundary for appropriate values of the normal derivative. This proves particularly advantageous when working in the field of blends and fillet generation, as has been discussed in earlier work [6].

The method of blend generation will now be demonstrated by way of an example.

Example

In Chapter 6.2 we saw that the surface produced in the example described a blend between a cylinder of unit radius centred on the z-axis and the flat plane given by $z = 0$. A blend of this nature may often be required inside a ship's hull where beams are to be fitted. We shall show in the next section that this blend is actually produced by our boundary-value approach to blend generation by solving Eq. 284 with suitable boundary conditions.

The domain of the solution is $0 \le u \le 1$, $0 \le v \le 2\pi$ with the trimlines defined on the cylinder by

$$\mathbf{r}(0, v) = (\cos v, \sin v, h) \tag{285}$$

and on the plane by

$$\mathbf{r}(1, v) = (R \cos v, R \sin v, 0). \tag{286}$$

Note that due to the range over which v varies and the sinusoidal nature of the boundary conditions, a closed loop of surface will be formed and conditions need only be imposed on the trimlines $u = 0$ and $u = 1$, as illustrated in Fig. 85b.

Since the equation being solved is fourth-order, we need impose the normal derivative of the function $\mathbf{r}$ as a boundary condition. If the boundaries of the domain are iso
u or v lines, these derivative boundary conditions are equivalent to specifying $\mathbf{r}_v$ on
a boundary along which u is constant and $\mathbf{r}_u$ on a boundary on which v is constant.
Then the function boundary condition and the derivatives effectively give us both $\mathbf{r}_u$
and $\mathbf{r}_v$ on the boundaries and hence the surface normal:

$$\mathbf{n} = \frac{\mathbf{r}_u \times \mathbf{r}_v}{|\mathbf{r}_u \times \mathbf{r}_v|}. \tag{287}$$

In blend generation the requirement of tangent plane continuity tells us what $\mathbf{n}$
must be on the boundary if the blend is to meet the primary surface smoothly; we can
thus use the above consideration to deduce derivative conditions that are consistent
with $\mathbf{n}$.

In this case a suitable set of tangency conditions are

$$\begin{aligned}
\mathbf{r}_u(0,v) &= (0,0,s) \\
\mathbf{r}_u(1,v) &= (b\sinh b\cos v, b\sinh b\sin v, 0)
\end{aligned} \tag{288}$$

where it should be noted that on each trimline $\mathbf{r}_u \times \mathbf{r}_v$ is parallel to the surface normal
on the neighbouring primary surface.

Hence we have a well posed boundary-value problem from which we can solve for
each of the independent variables (x, y, z) by considering Eq. 284.

Solution

The solution to this elliptic equation can either be sought analytically or numerically. The many methods for numerical solutions will be considered later, but for
now the solution to the above problem can be found analytically due to the nature
of the boundary conditions. The form of these suggest that we seek a solution that
contains terms proportional to $\cos v$ and $\sin v$.

Firstly, it should be noted that a solution which is proportional to $\cos v$ has the
form $F(u)\cos v$ where

$$\left(\frac{d^2}{du^2} - a^2\right)^2 F = 0 \tag{289}$$

i.e.

$$F \sim a_1 e^{au} + a_2 u e^{au} + a_3 e^{-au} + a_4 u e^{-au} \tag{290}$$

where a_1, a_2, a_3, a_4 are constants. A similar result is true for a solution proportional
to $\sin v$, and so we may deduce that the x and y components of the surface are of the
form:

$$\begin{aligned}
x &= (a_1 e^{au} + a_2 u e^{au} + a_3 e^{-au} + a_4 u e^{-au})\cos v \\
y &= (b_1 e^{au} + b_2 u e^{au} + b_3 e^{-au} + b_4 u e^{-au})\sin v
\end{aligned} \tag{291}$$

which, since $\cosh u = 1/2(e^u + e^{-u})$, $\sinh u = 1/2(e^u - e^{-u})$, can be written:

$$
\begin{aligned}
x &= ((c_1 + c_3 u) \cosh (au) + (c_2 + c_4 u) \sinh (au)) \cos v \\
y &= ((d_1 + d_3 u) \cosh (au) + (d_2 + d_4 u) \sinh (au)) \sin v.
\end{aligned}
\tag{292}
$$

The constants $c_1, \cdots, d_4$ are obtained by comparison with the boundary conditions on x and y.

The boundary conditions on z are given by

$$
\begin{array}{ll}
z(0) = h & z(1) = 0 \\
z_u(0) = s & z_u(1) = 0
\end{array}
\tag{293}
$$

from which we can see that the z component will be expressed as a polynomial of the form

$$
z(u) = z_0 + z_1 u + z_2 u^2 + z_3 u^3.
\tag{294}
$$

Hence by substitution of Eq. 293 into Eq. 294 we obtain

$$
z_0 = h, \quad z_1 = s, \quad z_2 = -3h - 2s, \quad z_3 = 2h + s
\tag{295}
$$

which gives

$$
\begin{aligned}
z &= h + su - (3h + 2s)u^2 + (2h + s)u^3 \\
&= (s + 3h)(u - 1)^2 + (s + 2h)(u - 1)^3.
\end{aligned}
\tag{296}
$$

Note that Eq. 292 and Eq. 296 give the same set of equations governing the surface blend in E^3 as in Chapter 6.2.

Example: Blend between a propeller blade and the hub

Consider now the generation of a blend between the central hub (a cylinder in this case) and the base of a propeller blade. This task is commonly used to make the machining of the surface easier and to strengthen the intersection between the blade and the hub. The fillet may be set up as a boundary-value problem as illustrated in Fig. 86.

The hub is of radius R_c, with the trimline on the hub of radius R_{bot} when projected onto the horizontal plane. The height of the blend is given by r_h at the midsection of the airfoil. The parameter domain over which the blend is generated is $0 \le u \le 1$, $-\pi/2 \le v \le \pi/2$ where a closed loop of surface is formed. The conditions on $u = 0$ are those of a generic propeller blade section:

$$
\begin{aligned}
x(0, v) &= (c \cos \beta + m_x \sin \beta) \cos v + t_x \sin \beta \sin 2v - \frac{m_x}{2}(\cos 2v + 1) \sin \beta \\
y(0, v) &= t_x \cos \beta \sin 2v + (m_x \cos \beta - c \sin \beta) \cos v - \frac{m_x}{2}(\cos 2v + 1) \cos \beta \\
z(0, v) &= R_c + r_h
\end{aligned}
\tag{297}
$$

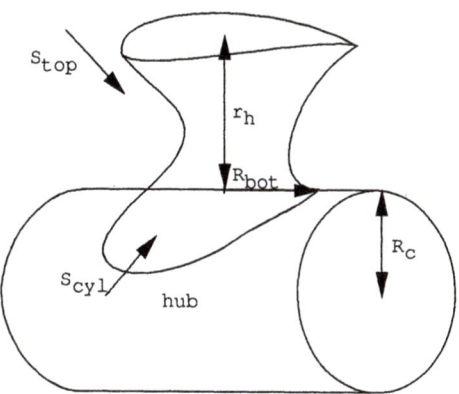

Fig. 86. The fillet geometry

where c defines the length of the blade, t_x the maximum thickness, m_x the maximum camber, and β the pitch of the blade section.

These equations representing the blade section will be explained more fully in the section on free-form design. For now they are taken as read.

On the cylinder we have a trimline enscribed on the surface of the cylinder whose projection onto the (x, y) plane is circular, as can be seen from Fig. 87. The x and y components can thus be defined by:

$$
\begin{aligned}
x(1, v) &= R_{bot} \cos 2v \\
y(1, v) &= R_{bot} \sin 2v
\end{aligned}
\tag{298}
$$

where R_{bot} is the projected radius of the trimline on the cylinder and $\cos 2v$ is taken since v covers the parameter range $-\pi/2 \leq v \leq \pi/2$. From Fig. 87 we can see that the radius of the hub is defined by

$$
R_c^2 = z^2 + y^2
\tag{299}
$$

which implies that the z component on the trimline will be given by

$$
z(1, v) = (R_c^2 - y^2)^{1/2} = (R_c^2 - R_{bot}^2 \sin^2 2v)^{1/2}.
\tag{300}
$$

The tangency conditions at the trimline are given by

$$
\begin{aligned}
x_u(0, v) &= S_x \cos 2v \\
y_u(0, v) &= S_y \sin 2v \\
z_u(0, v) &= S_{top}
\end{aligned}
\tag{301}
$$

in line with the conditions given in Section 7.

To define derivatives on the cylinder requires more work. Considering Fig. 87 we can see that the u-isoline will propagate radially. If the outer trimline is $u = 1$ then

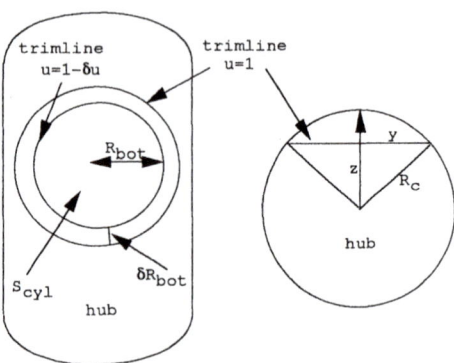

Fig. 87. The isolines on the hub

moving slightly inwards, a distance δu, gives the next trimline. We can define the parameter S_{cyl} to describe the rate at which the u-isolines propagate. Hence, x_u, y_u and z_u are given by S_{cyl} times a unit vector in the radial direction R_{bot}. Now:

$$\frac{\partial x}{\partial R_{bot}} = \cos 2v$$
$$\frac{\partial y}{\partial R_{bot}} = \sin 2v$$
$$\frac{\partial z}{\partial R_{bot}} = \frac{-R_{bot}\sin^2 2v}{(R_c^2 - R_{bot}^2\sin^2 2v)^{1/2}} \tag{302}$$

and to obtain a unit vector we divide each of the above quantities by

$$\left(\left(\frac{\partial x}{\partial R_{bot}}\right)^2 + \left(\frac{\partial y}{\partial R_{bot}}\right)^2 + \left(\frac{\partial z}{\partial R_{bot}}\right)^2 \right)^{1/2}$$
$$= \left(\cos^2 2v + \sin^2 2v + \frac{R_{bot}^2\sin^4 2v}{R_c^2 - R_{bot}^2\sin^2 2v} \right)^{1/2}$$
$$= \left(\frac{R_c^2 - R_{bot}^2\sin^2 2v + R_{bot}^2\sin^4 2v}{R_c^2 - R_{bot}^2\sin^2 2v} \right)^{1/2} \tag{303}$$

Hence, after manipulation of the equations we obtain the derivatives:

$$x_u(1,v) = S_{cyl}(1-t^2)^{1/2}\cos 2v$$
$$y_u(1,v) = S_{cyl}(1-t^2)^{1/2}\sin 2v$$
$$z_u(1,v) = -S_{cyl}t \tag{304}$$

where

$$t = \frac{R_{bot}\sin^2 2v}{(R_c^2 - R_{bot}^2\sin^2 2v + R_{bot}^2\sin^4 2v)^{1/2}}. \tag{305}$$

The generated fillet is illustrated in Fig. 88. It should be noted that the solution to this is obtained numerically due to the complex nature of the boundary conditions. For further details of this, and other problems, the reader is referred to [21].

Fig. 88. The generated fillet

3.1. Effect of Surface Control

There are two main areas of control of the generated surface, namely the derivative conditions and the smoothing parameter a. For tangent continuity it is required that $r_u \times r_v$ is parallel to the normal to the primary surface. It has been seen that r_v will be determined by the parametrization of the trimline on the primary surface. However, within the constraints of tangent continuity, there is still scope for modification of the surface through the choice of r_u at the boundary.

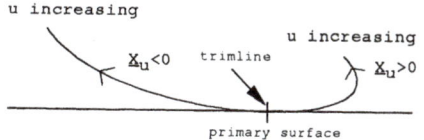

Fig. 89. Direction of approach to the primary surface

The direction of r_u determines the direction of approach of the blended surface to the trimline, while the magnitude of r_v determines the 'speed' of the isoparametric v-lines as they approach the trimline.

As can be seen from Fig. 89 changing the sign of r_u means reversing the direction in which the surface approaches the trimline. The magnitude of the derivative determines the speed of approach of the surface to the trimline.

If we consider the first example of the cylinder to plane blend where the z component is given by

$$z = (s + 3h)(u - 1)^2 + (s + 2h)(u - 1)^3, \tag{306}$$

we can consider the change in the surface produced by changing the values of s, the derivative magnitude. In Fig. 90b we can see that by increasing the value to $s = 10.0$ from $s = 2.0$ in Fig. 90a, the blend surface is pushed further away from the top trimline. This can be observed from the spacing of the lines of constant u.

(a)   (b)

Fig. 90. Alteration of the tangent magnitudes

Consider now a situation in which the slope parameter is a function of v. If we replace the condition that $z_u = s$ at $u = 0$ by

$$z_u = s_1 + s_2 \cos 3v \tag{307}$$

where s_1, s_2 are constants we now require that

$$z = z_0(u) + z_3(u) \cos 3v \tag{308}$$

where $z_0(u)$ is of the form of Eq. 306 and $z_3(u)$ is of the form of Eq. 290 with the constants a_1, a_2, a_3, a_4 determined by

$$\begin{array}{ll} z_3(0) = 0 & z_3(1) = 0 \\ z_{3u}(0) = s_2 & z_{3u}(1) = 0. \end{array} \tag{309}$$

Fig. 91a shows the case where $s_1 = 1, s_2 = 2$. This now gives a surface in which ridges are present which can be used to add strength to the blend. Furthermore, it can be shown that for large values of a, the solution for $z_3(u)$ can be approximated by

$$z_3(u) \sim s_2 u e^{-3au}. \tag{310}$$

which indicates the presence of a thin layer near $u = 0$ where the effects of the sinusoidal variation in the derivative boundary conditions is restricted to a thin layer. This is illustrated in Fig. 91a, where $a = 1.8$. For low values of a the influence of the boundary conditions at $u = 0$ extends throughout the whole region as in Fig. 91b where $a = 0.5$.

(a)   (b)

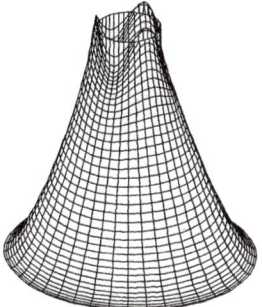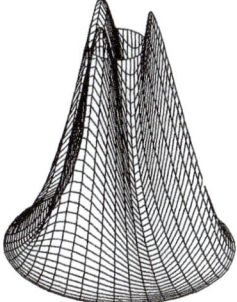

Fig. 91. A ridged surface for varying values of a

3.2. The Smoothing Parameter a

The parameter a controls the relative smoothing of the dependent variables between the u and v directions. It can take many forms, such as being a function of v, a delta function or take a constant value. It can be seen for the case where a is constant that by replacing the partial differential operator by

$$\left(\frac{\partial^2}{\partial u^2} + \frac{\partial^2}{\partial (v/a)^2} \right) \tag{311}$$

that the parameter a changes the length scale in the v direction.

The effect of the boundary layer can be seen by considering the following illustrative example (Fig. 92):

$$\epsilon \frac{dy}{dt} + y = 1 \tag{312}$$

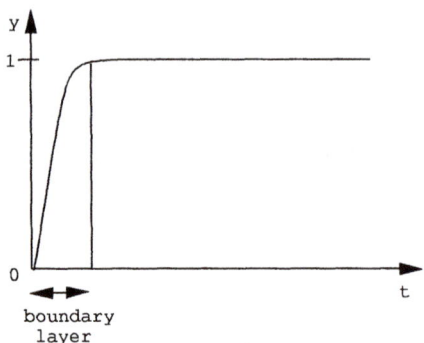

Fig. 92. The boundary layer solution

where $\epsilon > 0$ and $\epsilon \ll 1$, and boundary conditions are such that $y = 0$ at $t = 0$.

It is tempting to think that a good approximation to the above equation is $y = 1$ since ϵ is small, however this implies that the boundary condition at $t = 0$ is never satisfied, no matter how small ϵ. The full solution is given by

$$y = 1 - e^{-\frac{t}{\epsilon}}. \tag{313}$$

A small value of ϵ is compensated by a large dy/dt in the boundary layer which is of thickness $O(\epsilon)$ in t. Therefore in the PDE method it can correspondingly be seen that the boundary layer is of thickness $O(1/a)$ in u, which means that changes in the u direction occur over a relatively short length scale for large a and over a large scale for small a.

Cheng discusses further the smoothing parameter a [15]. He maintains that the smoothing parameter can be thought of as a 'thumbweight'. Therefore for small values of a, a 'fuller' surface will be generated, while for larger values a blend will be generated which has a thinner 'waistline'. This can be seen below in which the values of the smoothing parameter are altered from $a = 0.2$ in Fig. 93a to $a = 2.0$ in Fig. 93b.

Example

Consider a blend from a fluted cylinder to the plane $z = 0$. The trimlines are given by

$$
\begin{aligned}
x &= \cos v + \epsilon \cos 4v \\
y &= \sin v - \epsilon \sin 4v \\
z &= H
\end{aligned}
\tag{314}
$$

on the cylinder, and

$$x = R \cos v + \delta \cos 4v$$

(a)   (b)

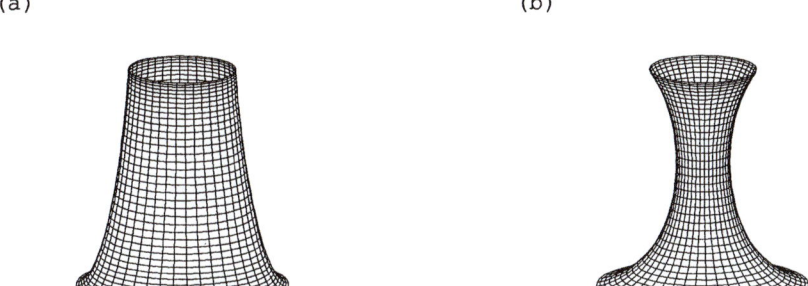

Fig. 93. The influence of the smoothing parameter a

$$
\begin{aligned}
y &= R\sin v - \delta \sin 4v \\
z &= 0
\end{aligned}
$$
(315)

on the plane. We will look for a solution of the form

$$
\begin{aligned}
x &= \cosh(a_0 u)\cos v + X_4(u)\cos 4v \\
y &= \cosh(a_0 u)\sin v + Y_4(u)\sin 4v \\
z &= \text{cubic in } u
\end{aligned}
$$
(316)

where X_4, Y_4 are of the form

$$
X_4 = A_1 e^{\sigma u} + A_2 u e^{\sigma u} + A_3 e^{-\sigma u} + A_4 u e^{-\sigma u}
$$
(317)

with $\sigma = 4a$.

To illustrate the point about the boundary layer further, it can be shown that for large values of σ, an approximation to X_4 is given by

$$
X_4 \approx (\sigma\delta + \delta - s)e^{\sigma(u-1)} + (t - \sigma\delta)u e^{\sigma(u-1)} + \epsilon e^{-\sigma u} + \sigma\epsilon u e^{-\sigma u}
$$
(318)

where s is assumed to be a constant tangent magnitude; Eq. 318, which indicates the presence of a thin layer of thickness $O(1/\sigma)$ near $u = 0$ and $u = 1$ where the effects of the variations in ϵ and δ are felt, but beyond which the cross-section is circular, as can be seen in Fig. 94.

Therefore we have a general method for generating blends between primary surfaces. The wide scope for surface control of the blend surface has been illustrated, and this will be of significance in the next section where we deal with the concepts of free-form surface design using the PDE method.

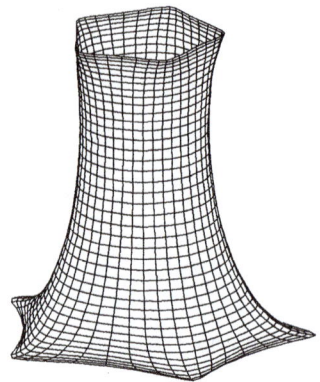

Fig. 94. The fluted cylinder blend

FREE-FORM SURFACE DESIGN

4. Introduction

The defining characteristic of a free-form surface is that it can be changeable by the designer. It is convenient to represent free-form surfaces in terms of surface patches, often given as Bézier and B-spline surfaces. The PDE method can easily be extended from blend design into the area of free-form design by relaxation of the continuity conditions on the boundaries [8]. In the previous section it was required that continuity be preserved across the boundary between primary and secondary surfaces; free-form design adds a new dimension to surface manipulation, in that the derivative conditions are now free to be used to control the surface, with the position of the boundary curves being left to the designer to define and not rely upon any prespecified trimlines. This section will deal with the way in which free-form surfaces can be generated, by way of several examples.

5. Free-Form Surfaces from Partial Differential Equations

The surface is again sought as the solution to the modified biharmonic operator, given by Eq. 284 in Section 3. It was seen in Section 3 that an analytic solution could be obtained which is proportional to $\cos(v)$ with the form $F(u)\cos(v)$ where $F(u)$ is given by Eq. 290. In general, an analytic solution to a problem with periodic boundary conditions can be obtained if the boundary conditions are given in terms of cosine and sine functions of v. If this is the case, the general solution will be of the form:

$$r(u,v) = A_0(u) + \sum_{n=1}^{\infty} A_n(u)\cos nv + B_n(u)\sin nv \qquad (319)$$

where

$$A_0 = a_{00} + a_{01}u + a_{02}u^2 + a_{03}u^3, \qquad (320)$$

$$\begin{aligned} A_n &= a_{n1}e^{anu} + a_{n2}ue^{anu} + a_{n3}e^{-anu} + a_{n4}ue^{-anu} \\ B_n &= b_{n1}e^{anu} + b_{n2}ue^{anu} + b_{n3}e^{-anu} + b_{n4}ue^{-anu} \end{aligned} \qquad (321)$$

and where $a_{n1}, \cdots, a_{n4}, b_{n1}, \cdots, b_{n4}$ are vector-valued constants which are determined by the boundary conditions imposed on the isoparametric lines forming the boundaries. It is clear that if this method is to be used as a design tool, an understanding of the way in which the properties, boundary conditions and parameters of the PDE affect the geometric properties of the surface is needed.

If we consider the boundary conditions to be simplified of the form:

$$\begin{aligned} r(0,v) &= p_0 & r(1,v) &= p_1 \\ r_u(0,v) &= s_0 & r_u(1,v) &= s_1 \end{aligned} \qquad (322)$$

where $\mathbf{p}_0, \mathbf{p}_1, \mathbf{s}_0, \mathbf{s}_1$ are vector constants, then the only term from the general solution of Eq. 319 will be given by the term $\mathbf{A}_0(u)$. This represents a cubic polynomial curve in the parameter u that lies between the points with position vectors $\mathbf{p}_0$ and $\mathbf{p}_1$, with the direction and 'speed' of the curve at these points determined by $\mathbf{s}_0$ and $\mathbf{s}_1$. This polynomial curve can be regarded as the spine of the surface.

If we return to the general form of the boundary conditions then the Fourier coefficient functions $\mathbf{A}_n$ and $\mathbf{B}_n$ will, in general, no longer be zero. It should be noted that the decaying and growing exponential functions within $\mathbf{A}_n$ and $\mathbf{B}_n$ result in the smooth transition between the boundary conditions on $u = 0$ and $u = 1$. It should further be noted that the rate at which the exponentials grow and decay is determined by the smoothing parameter a, and in particular the factor an. The higher the frequency modes, n, in the boundary conditions, the more rapidly these terms are smoothed out.

Thus, if only the first Fourier mode is present such that:

$$\mathbf{r}(u,v) = \mathbf{A}_0(v) + \mathbf{A}_1(u)\cos v + \mathbf{B}_1 \sin v \qquad (323)$$

then this will represent a solution in which the boundary curves $u = 0$ and $u = 1$ are general conic sections, each centred on the point $\mathbf{p}_0$ and $\mathbf{p}_1$ respectively. If we recall that $\mathbf{A}_0(u)$ represents a cubic spine between the points $\mathbf{A}_0(0) = \mathbf{p}_0$ and $\mathbf{A}_0(1) = \mathbf{p}_1$ then we may regard the surface as being formed by a conic that is swept through space between these end points whose centre lies on the spine. The shape of the conic is dependent on u and so transforms from the shape at the boundary curve $u = 0$ to that at $u = 1$.

If we generalise the boundary conditions such that the solution is given by Eq. 319, we can now see that the vector surface $\mathbf{r}(u,v)$ can be regarded as being composed of the spine vector $\mathbf{A}_0(u)$ plus a primary radius vector $\mathbf{A}_1(u)\cos v + \mathbf{B}_1(u)\sin v$ plus a secondary radius vector $\mathbf{A}_2(u)\cos 2v + \mathbf{B}_2(u)\sin 2v$ that is attached to the end of the primary radius and so forth.

This can be thought of as a simplified model to picture the surface, however it should be borne in mind that this assumes that the amplitude of the Fourier coefficient functions decay with increasing frequency.

The boundary conditions on the function $\mathbf{r}$ are chosen so that the curves from the edges of the surface patch have the desired shape. With reference to the section on blend design, the parameter v will vary over some range in order to render a closed surface and so that no boundary conditions are required on the lines of constant v, but solely on the trimlines given by $u = 0$ and $u = 1$. To get a feel for the way in which the boundary values of $\mathbf{r}_u$ and $\mathbf{r}_v$ affect the shape, it should be remembered that the direction of the vectors are tangential to the isoparametric u and v lines of the surface and the magnitudes of these vectors gives the 'speed' of the curves. Therefore, by altering the values specified for $\mathbf{r}_u$ and $\mathbf{r}_v$ along the boundaries, the direction in which the surface moves away from the edges of the patch, and how far it moves in space before it begins to respond to the boundary conditions on other parts of the patch can be affected.

It should be noted at this stage that the closed surface offers a wide scope for possible surface shape in the design of free-form surfaces, which will be illustrated by way of several examples.

6. A Simple Hull Design

Consider now the problem of creating a hull of a ship or boat using a single periodic patch. At this stage it is constructive to consider an analytic solution by choosing suitable boundary conditions to illustrate the influence of design parameters and boundary conditions. It should be noted that a numerical solution offers a great deal more scope to the designer, particularly in terms of local control in the choice of boundary conditions, and this will be considered later.

6.1. Initialization of Problem

We need to set up the problem as a boundary-value problem in (u, v) space with boundary conditions specified along curves in the (u, v) plane which correspond to closed curves in E^3. We therefore take one of the boundary curves to be the plan outline of the hull at the deck level. On the curve, u is taken as zero and the shape is given parametrically in terms of v. The Cartesian coordinate system is then determined such that x is measured along the length of the deck, y is measured from port to starboard, and z is taken as the hull level, as illustrated in Fig. 95.

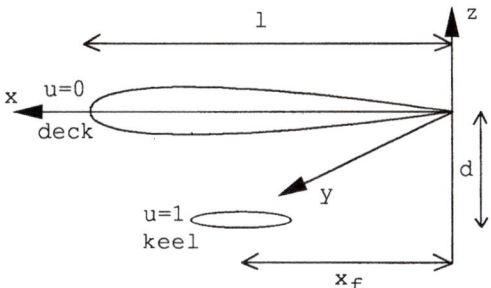

Fig. 95. The trimlines for the yacht geometry

Then, since the deck geometry is required to be such that a blunt stern and sharp bow are required (as is the profile curve in Fig. 95) the boundary curve is defined by

$$
\begin{aligned}
x(0, v) &= l \cos v \\
y(0, v) &= \sin 2v + b_4 \sin 4v + b_6 \sin 6v \\
z(0, v) &= 0
\end{aligned}
\tag{324}
$$

over the parameter range $-\pi/2 \leq v \leq \pi/2$, so that the singular nature of the surface at the tip of the bow can be accounted for in the closed-form solution.

For a yacht like hull, the conditions on $u = 1$ we take to be the bottom of the keel with an elliptic plan section, and so the boundary conditions are given by

$$
\begin{aligned}
x(1,v) &= x_f + a_f \cos 2v \\
y(1,v) &= b_f \sin 2v \\
z(1,v) &= -d
\end{aligned}
\tag{325}
$$

where d is the depth of the keel, and x_f represents the position of the centreline of the keel, relative to the deck. For this particular surface, the various constants are given by: $b_1 = 6, b_4 = -0.1, b_6 = -0.06, d = 1.5, x_f = 3.7, a_f = 0.5, b_f = 0.05$.

The boundary conditions on (x, y, z) are determined and so it remains to specify the derivatives on the boundaries. In the case of blend generation only a limited amount of freedom was available to the designer in choosing these conditions to influence the design (only the vector $\mathbf{r}_u$ could be altered, as $\mathbf{r}_v$ was fixed by the parametrization of the trimlines). In contrast, in the present case the specification of the derivative boundary conditions is a major part of the design process. The way in which appropriate values and distributions can be determined is intuitively straightforward in this example.

If we impose the conditions at the deck level:

$$
\begin{aligned}
x_u(0,v) &= -S_{X_0} + S_{X_1} \cos v \\
y_u(0,v) &= S_{Y_0} \sin 2v \\
z_u(0,v) &= -S_{Z_0}
\end{aligned}
\tag{326}
$$

where $S_{X_0}, S_{X_1}, S_{Y_0}$ and S_{Z_0} are parameters, then we have control over the way in which the isoparametric v lines leave the boundary curve $u = 0$ (or deck).

If we take the parameters S_{X_1} and S_{Y_0} to be zero then the conditions will imply that the isoparametric v lines leave the boundary curve $u = 0$ at an angle of $\tan^{-1}(S_{X_0}/S_{Z_0})$ and the bows will be raked backwards to produce a retroussé stern.

On $u = 1$, the boundary curve at the bottom of the boat, to obtain a yacht-like hull, we take

$$
\begin{aligned}
x_u(1,v) &= -S_{X_f} \\
y_u(1,v) &= 0 \\
z_u(1,v) &= -S_{Z_f}
\end{aligned}
\tag{327}
$$

which ensures that a fin like shape for the keel is obtained, providing S_{Z_f} is sufficiently large and positive; and further the keel is raked towards the stern at an angle of $\tan^{-1}(S_{X_f}/S_{Z_f})$ to the vertical.

Fig. 96. A yacht with full hull

By making the parameter S_{Y_0} at the deck non-zero, we can influence the 'fullness' of the hull. The greater the value of S_{Y_0}, the more the body of the vessel will bulge out beyond the deck outline. This can be seen from Fig. 96 where the value of S_{Y_0} is taken to be 2, and $S_{X_0} = 2, S_{Z_0} = 3, S_{X_f} = 2, S_{Z_f} = 4.5$.

In order to modify the shape of the stern we can alter the parameter S_{X_1}. It is clear that the retroussé stern is a consequence of the rake effect introduced by the conditions on $r_u(0, v)$. If the parameter S_{X_0} is taken to be zero, the resulting shape will show no rake at the deck. To retain the rake at the bows and on the fin but reverse the slope of the surface as it approaches the deck at the stern, the parameter S_{X_1} is altered, in the case of Fig. 97, where it is given by $S_{X_1} = 3$.

Fig. 97. A yacht with reversed slope at the deck level

Furthermore, the smoothing parameter a influences the general shape of the hull. To obtain a gradual transition from the conditions imposed at the deck to those imposed at the fin, a small value of a_x is chosen. If a_y is also taken to be small then the 'bulge' in the yacht will be less pronounced and the yacht will be more slender.

The solution for the yacht is consequently obtained from the chosen boundary conditions to give the analytic form of the yacht as

$$
\begin{aligned}
x(u,v) &= x(u) + X1\cos v + X2\cos 2v \\
y(u,v) &= Y2\sin 2v + Y4\sin 4v + Y6\sin 6v \\
z(u,v) &= z(u)
\end{aligned}
\tag{328}
$$

where $x(u), z(u)$ are polynomials of the form of Eq. 320 and $X1, X2, Y2, Y4, Y6$ are of the form of Eq. 321. Thus, we see that an analytic solution is obtained almost instantaneously, making the generation and manipulation of the PDE surface very fast and efficient.

This example serves to illustrate the control in the resulting surface which is exercised by the boundary conditions and smoothing parameter. The next example we will consider is that of propeller blade design.

7. Propeller Blade Design

A propeller blade is another good example of a surface which can be generated using the PDE method. A propeller blade geometry is often supplied as two dimensional data in the form of blade sections from which three dimensional coordinates are generated. This is explicitly illustrated in Chapter 8 in the example of the skinning method. We therefore have a good opportunity to compare the surface generation technique of the PDE method with that of B-spline surface generation.

7.1. Initialization of Problem

As in the case of the generic yacht we may generate the surface as a single closed loop of surface parametrized by $-\pi/2 \le v \le \pi/2$, $0 \le u \le 1$. The trimlines are taken on $u = 1$ to be the cross-sectional profile of the blade at the base, with the trimline $u = 0$ corresponding to the tip of the propeller. The parametrization we obtain for the blade section is given by Eq. 297 in Section 3 of this chapter.

The cross-section profile throughout the blade is commonly of a type of aircraft wing section with a thickness and camber distribution (see Fig. 122, Chapter 8). To parametrize this generically, we may take

$$
\begin{aligned}
x &= c\cos v \\
y &= t_x\sin 2v
\end{aligned}
\tag{329}
$$

which gives a trimline similar to that of the yacht of length c, maximum thickness t_x. To this profile we add a camber line:

$$
\begin{aligned}
x &= c\cos v \\
y &= t_x\sin 2v + m_x\left(\cos v - \frac{\cos 2v + 1}{2}\right)
\end{aligned}
\tag{330}
$$

This we then rotate to an appropriate pitch angle β with the standard transformation:

$$\begin{aligned} x' &= x\cos\beta + y\sin\beta \\ y' &= y\cos\beta - x\sin\beta \end{aligned} \qquad (331)$$

to give the parametrization described by Eq. 297. for the fillet design. The parameter β gives the twist, or pitch of the blade, and is effectively used to align the angle of each section profile to the incoming flow, in order that a required amount of lift is generated.

The trimline at the tip $(u = 0)$ is formed by

$$\begin{aligned} x(0,v) &= f \\ y(0,v) &= 0 \\ z(0,v) &= d. \end{aligned} \qquad (332)$$

The parameter f is an offset along the x-axis and gives a measure of the skew of the blade. The boundary-value problem is illustrated in Fig. 98.

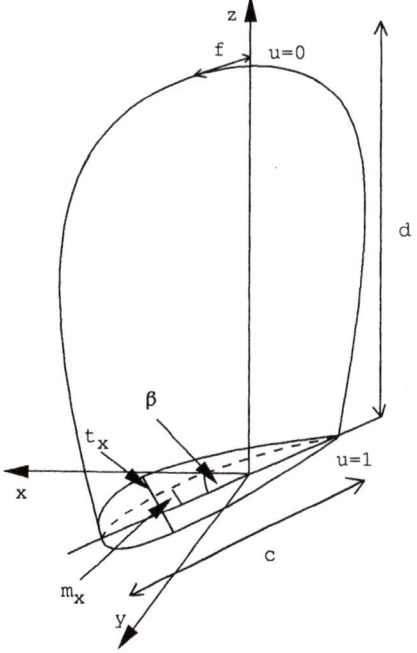

Fig. 98. The trimlines for the propeller blade

In a similar fashion to the case of the yacht hull, the tangency conditions are given at the blade tip by

$$
\begin{aligned}
x_u(0,v) &= S_x \cos 2v \\
y_u(0,v) &= S_y \sin 2v \\
z_u(0,v) &= 0
\end{aligned}
\tag{333}
$$

and at the blade root by

$$
\begin{aligned}
x_u(1,v) &= E_x \cos 2v \\
y_u(1,v) &= E_y \sin 2v \\
z_u(1,v) &= S_{bot}.
\end{aligned}
\tag{334}
$$

For a marine propeller, since a smooth curve is required at the tip for the blade profile outline, the derivative z_u is set equal to zero.

From the governing boundary conditions the marine propeller can be generated by solving the elliptic PDE (Eq. 284) to give the solution for each of the blades as

$$
\begin{aligned}
x(u,v) &= x(u) + X1 \cos v + X2 \cos 2v + XS2 \sin 2v \\
y(u,v) &= y(u) + Y2 \sin 2v + YC1 \cos v + YC2 \cos 2v \\
z(u,v) &= z(u)
\end{aligned}
\tag{335}
$$

where the functions $x(u), X1, \cdots, YC2$ are defined by Eqs. 320 and 321. Note that the surface is C^0 and C^1 continuous between the blade and the fillet of Section 3.

7.2. Actual Blade Representation

For an actual blade we must parametrize the governing equations used to define the section. Thus, in Chapter 8, we have the equations for the thickness distribution, meanline distribution and section coordinates given by Eqs. 444, 445 and 448/449 respectively. Each of these we must Fourier analyse to provide a parametrization consistent with the analytic solution of the PDE. Instead of Eq. 297 representing the parametrization of the section at the base, we now have:

$$
\begin{aligned}
x(1,v) &= X_f(v) + Y_{tsf}(v) \\
y(1,v) &= Y_{mf}(v) + Y_{tcf}(v)
\end{aligned}
\tag{336}
$$

where Y_{mf}, Y_{stf} correspond to the Fourier analysed components $y_m, y_t \sin \theta$, etc. of Eqs. 448 and 449. It should be noted that up to only four terms are required in the Fourier representation.

Control of the blade is important since we require the section profile to be maintained throughout the blade span. To achieve this, and to obtain control of each of the thickness, mean line, etc. distributions we impose first and second derivatives at the base and tip of the form:

$$
\begin{aligned}
x_u(1,v) &= S_c X_f(v) + S_t Y_{tsf}(v) & x_{uu}(1,v) &= C_c X_f(v) + C_t Y_{tsf}(v) \\
y_u(1,v) &= S_m Y_{mf}(v) + S_t Y_{tcf}(v) & y_{uu}(1,v) &= C_m Y_{mf}(v) + C_t Y_{tcf}(v)
\end{aligned}
\tag{337}
$$

where S_c etc. are parameters controlling the first derivative terms, and C_c etc. control the second derivatives. It should be emphasized that the elliptic eqn (Eq. 284) is now a sixth order PDE since we require second derivatives. This parametrization provides an almost exact interpolation of a propeller geometry, with the blade section profile maintained throughout. Furthermore, the parameters ensure we have independent control of each of the distributions along the blade span. The sections must finally be placed on cylindrical sections of the blade to obtain the final surface (as shown in Fig. 121, Chapter 8). This is achieved by taking each of the u-isolines and using a transformation of the type given by Eq. 451. Further details of this example and its use in design and analysis of propellers are given in [21]. Fig. 99 illustrates the PDE generated propeller.

Fig. 99. The marine propeller

8. The Telephone Handset

The final analytic example considers the generation of a telephone handset. This may be required so that a polymer cast can be produced from which to make multiple identical surfaces. In this example the handset is generated from four closed-form surface patches; two patches for the main body of the handset, and one patch for the mouthpiece and one for the earpiece.

To generate one of the two surface patches that represent the body of the handset the boundary conditions are imposed:

$$
\begin{aligned}
x(0,v) &= k + (r\sin\alpha)\cos v \\
y(0,v) &= r\sin v \\
z(0,v) &= h - (r\cos\alpha)\cos v
\end{aligned}
\tag{338}
$$

and

$$
\begin{aligned}
x(1,v) &= R\cos v \\
y(1,v) &= R\sin v \\
z(1,v) &= 0
\end{aligned}
\tag{339}
$$

where the curve on $u = 0$ corresponds to a circle of radius r centred upon the point $\mathbf{r} = (k, 0, h)$ while the curve $u = 1$ is a circle of radius R centred on the origin.

The derivative conditions are obtained from the directions of the surface normals $\mathbf{n}(u, v)$ at the isoparametric lines $u = 0$ and $u = 1$ and are given by

$$
\begin{aligned}
x_u(0,v) &= -s_0\cos\alpha/r \\
y_u(0,v) &= 0 \\
z_u(0,v) &= -s_0\sin\alpha/r
\end{aligned}
\tag{340}
$$

and

$$
\begin{aligned}
x_u(1,v) &= s_1\cos\beta\cos v/R \\
y_u(1,v) &= s_1\cos\beta\sin v/R \\
z_u(1,v) &= -s_1\sin\beta/R
\end{aligned}
\tag{341}
$$

where the constant β can be varied to alter the shape of the receiver, and s_0 and s_1 are arbitrary constants.

The boundary conditions required to represent the mouthpiece/earpiece of the handset are considered next. On $u = 0$ the boundary conditions are given by

$$
\begin{aligned}
x(0,v) &= R\cos v \\
y(0,v) &= R\sin v \\
z(0,v) &= 0
\end{aligned}
\tag{342}
$$

and on $u = 1$

$$
\begin{aligned}
x(1,v) &= \epsilon\cos v \\
y(1,v) &= \epsilon\sin v \\
z(1,v) &= d
\end{aligned}
\tag{343}
$$

where $\epsilon \to 0$. The derivative boundary conditions are given by

$$
\begin{aligned}
x_u(0,v) &= s_3 \cos\beta \cos v / R \\
y_u(0,v) &= s_3 \cos\beta \sin v / R \\
z_u(0,v) &= -s_3 \sin\beta / R
\end{aligned}
\tag{344}
$$

and

$$
\begin{aligned}
x_u(1,v) &= s_4 \cos v \\
y_u(1,v) &= s_4 \sin v \\
z_u(1,v) &= 0
\end{aligned}
\tag{345}
$$

where s_3 and s_4 are taken to be constants and it should be noted from the tangency conditions given by Eq. 341 and Eq. 344, that there is continuity of tangent direction between the mouthpiece patch and the patch representing the body of the handset. Having obtained one half of the phone, the other half is obtained by joining it to two similar patches which represent the other half of the handset. Fig. 100 illustrates the final free-form surface.

Fig. 100. The four patched telephone receiver

9. Numerically Obtained Surfaces

Having considered what may be achieved using a simple periodic solution, let us now consider what may be achieved by using a solution to a numerically posed boundary-value problem. The boundary data has been chosen so as to give a surface with a particular shape, rather than a surface that must meet certain constraints such as blend generation. In this example we will consider the problem of generating a surface that resembles a ship hull, or in particular half a ship hull. We solve the PDE over the square region $0 \leq u \leq 1, 0 \leq v \leq 1$ of the (u, v) plane using a numerical solution technique, as will be described in the next section. Since the PDE is 4th order, boundary conditions on the coordinate vectors $\mathbf{r}_u$ and $\mathbf{r}_v$ are also required in keeping with the previous examples. The boundary conditions on $\mathbf{r}(u, v)$ were chosen so as to give a suitable outline for the profile of the ship's hull, those being the stem, stern, keel and deck. Since the data is imposed numerically, this consists of choosing points along the appropriate space curves. The mapping between points along the boundary of the (u, v) parameter domain and points in E^3 has been chosen so that the boundary curves of the hull have arc-length parametrization. There is no special reason for this, it is merely a convenient way of ensuring that the numerical resolution is adequate at all points across the hull's surface. The derivative boundary conditions were chosen so that the interior surface of the ship's hull would have a geometrically reasonable shape. In practise this could be achieved by considering the section curves and waterlines of the design. The conditions consisted in specifying $\mathbf{r}_u$ along the curves $\mathbf{r}(0, v)$ and $\mathbf{r}(1, v)$ and specifying $\mathbf{r}_v$ along the curves $\mathbf{r}(u, 0)$ and $\mathbf{r}(u, 1)$. It should be noted that a non-periodic surface patch is considered and so four sets of boundary data need to be specified. The magnitudes and directions of the coordinate vectors were varied until the desired shape was attained for the hull.

The final hull was created by joining together two identical hull parts to create the complete shell as in Fig. 101.

If a more accurate representation were required then the hull surface could be discretized into many surface patches, and fitted together by ensuring tangent plane continuity at the boundaries of the patches, as illustrated in Chapter 9. It should further be recalled that the PDE surface patches are smooth and so fairing of the ship design (which will be discussed later) is thus minimized.

Other areas of design have been looked into, particularly by Cheng [15], in which the bossing between a ship's hull and the shaft of a marine propeller was considered. The procedure for obtaining the necessary trimlines for the boundary-value problem consisted of taking the discrete data points on the ship's hull from a plan view of the ship's surface; the data for the other trimline was taken from the shaft running to the operating propeller. The surface was obtained numerically with an added degree of manipulation coming from a variable value of the smoothing parameter which could vary throughout the surface, due to the nature of the numerical solution. The design is illustrated in Fig. 102.

Finally, Fig. 103 illustrates a chine which is generated from two surface patches joined to form the complete surface.

Of course, to obtain these surfaces, and many other surfaces from a set of discrete boundary conditions, the numerical solutions to elliptic equations must be considered. These will be looked into in the next sections so that these surfaces can readily be obtained.

Fig. 101. The numerically obtained ship form

Fig. 102. The bossing between the hull and propeller hub

Fig. 103. A chine

NUMERICAL SOLUTIONS OF ELLIPTIC PDES

10. Introduction

We have generated surfaces as solutions of PDEs with suitably chosen boundary conditions. When these boundary conditions are given as a sum of Fourier modes, then an analytic solution can be sought by separation of variables, which results in the solution being derived in the form of Eq. 319. However there are many instances, such as those illustrated in the last section and when the boundary curves are not defined in closed-form, when a numerical solution is the only means of evaluating the solution.

Of the numerical methods available for solving differential equations, those employing finite differences [70] are among the most frequently used and applicable, and this section will deal in detail with such methods; it will then go on to deal with other methods of approximation - in particular collocation techniques [17].

11. Finite Difference Methods

Finite difference methods are approximations in the sense that derivatives at a point in the solution domain are approximated by difference quotients over a small interval. The solution to the elliptic equation is sought by discretizing the bounded domain by a rectangular mesh formed by the isoparametric lines given by u =const, v =const. An approximate solution to the PDE is found by estimating the value of the function at the points of intersection which are called the mesh points. The solution process itself consists of approximating the PDE at the mesh point $P_{i,j}$ say, by a finite difference approximation in terms of the solution value at that point $x_{i,j}$ and the solution values at neighbouring points. From this a system of linear equations is derived which can be solved to obtain the values at the mesh points.

The accuracy of these methods is determined by how finely the solution domain is discretized by the user, and also by the rounding-error of the computing hardware.

11.1. Finite Difference Approximations for Derivatives

If a function u and its derivatives are single valued, finite and continuous functions of x, then by using a Taylor expansion we obtain

$$
\begin{aligned}
x(u + h) &= x(u) + hx'(u) + 1/2h^2x''(u) + 1/6h^3x''' + \cdots \\
x(u - h) &= x(u) - hx'(u) + 1/2h^2x''(u) - 1/6h^3x''' + \cdots
\end{aligned}
\tag{346}
$$

Addition of these gives

$$
x(u + h) + x(u - h) = 2x(u) + h^2x''(u) + O(h^4)
\tag{347}
$$

where $O(h^4)$ denote the terms containing fourth and higher powers of h. Assuming these are negligible in comparison with terms containing lower powers of h, this gives

$$x''(u) \approx \frac{1}{h^2} \left[x(u+h) - 2x(u) + x(u-h) \right] + O(h^2). \tag{348}$$

Subtracting Eq. 346 gives

$$x'(u) \approx \frac{1}{2h} \left[x(u+h) - x(u-h) \right] + O(h^2) \tag{349}$$

with the discretization error of order h^2.

Eq. 349 is termed a central difference approximation and it approximates the slope of the tangent at u by the average of the slope at $u-h$ and the slope at $u+h$, as can be seen from Fig. 104. It should be noted that Eq. 348 gives the derivative of the gradient at the point C.

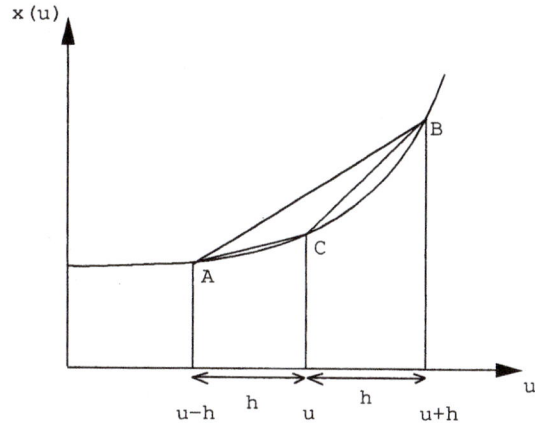

Fig. 104. Finite difference approximations to the slope of the curve

Alternative methods for approximating the slope of the tangent are given by the forward difference formula

$$x'(u) \approx \frac{1}{h} \left[x(u+h) - x(u) \right] + O(h) \tag{350}$$

or the backward difference formula

$$x'(u) \approx \frac{1}{h} \left[x(u) - x(u-h) \right] + O(h) \tag{351}$$

where both have leading errors of order $O(h)$. In general, for elliptic equations, it is better to use the central difference approximation to the tangent slope as the error is of a higher order.

11.2. Application to the Solution of the PDE Problem

Consider now the problem of solving the PDE

$$\left(\frac{\partial^2}{\partial u^2} + a^2 \frac{\partial^2}{\partial v^2} \right) \mathbf{r} = \mathbf{f}(u, v) \tag{352}$$

over the domain Ω in the (u, v) parameter space with suitable boundary conditions. The (u, v) parameter space can conveniently be divided into a mesh by taking the isoparametric lines $u =$ const, $v =$ const, where the mesh has a uniform grid of size

$$\begin{aligned} \delta u &= h \\ \delta v &= k \end{aligned} \tag{353}$$

where $h = (u_1 - u_0)/(p - 1)$, $k = (v_1 - v_0)/(q - 1)$ and $p * q = n$ is the total number of mesh points which comprise the grid.

Then, for a general point $\mathbf{P}_{i,j}$ in the domain of the mesh, the parametric coordinates will be given by $u = ih, v = jk$ where i and j are integers. Furthermore, if $\mathbf{r}$ represents a surface with cartesian coordinates given by $\mathbf{r} = (x, y, z)$, then the value of $\mathbf{r}$ at $\mathbf{P}_{i,j}$ will be denoted by

$$\mathbf{r}_{i,j} = (x_{i,j}, y_{i,j}, z_{i,j}). \tag{354}$$

To obtain the solution we need boundary conditions. In the case of a second order elliptic equation, the value of $\mathbf{r}_{i,j}$ can be specified at each point which lies on the boundary isolines, $u = u_0, u = u_1, v = v_0, v = v_1$, in terms of the parameters u and v. This then implies that if the mesh has $p \cdot q$ mesh points, then there will be $(p - 2) \cdot (q - 2)$ unknown values of $\mathbf{r}_{i,j}$ inside the domain of the parameter space.

If we now consider the x component of the PDE given by Eq. 352 over the domain $0 \leq u \leq 1, 0 \leq v \leq 2\pi$ in the (u, v) parameter space, then from the finite difference approximations for the second derivatives, we obtain at the mesh point $\mathbf{P}_{i,j}$:

$$\left(\frac{\partial^2 x}{\partial u^2} \right)_{i,j} \approx \frac{x_{i+1,j} - 2x_{i,j} + x_{i-1,j}}{h^2} \tag{355}$$

and

$$\left(\frac{\partial^2 x}{\partial v^2} \right)_{i,j} \approx \frac{x_{i,j+1} - 2x_{i,j} + x_{i,j-1}}{k^2} \tag{356}$$

and the PDE can be approximated by the finite difference scheme

$$\frac{x_{i+1,j} - 2x_{i,j} + x_{i-1,j}}{h^2} + a^2 \left(\frac{x_{i,j+1} - 2x_{i,j} + x_{i,j-1}}{k^2} \right) = f_{i,j} \tag{357}$$

which can be conveniently represented in the form

$$\begin{bmatrix} & r^2 & \\ 1 & -2(1 + r^2) & 1 \\ & r^2 & \end{bmatrix} x_{i,j} = h^2 f_{i,j} \quad \begin{matrix} i = 1, \cdots p \\ j = 1, \cdots q \end{matrix} \tag{358}$$

where

$$r^2 = \frac{a^2 h^2}{k^2} \tag{359}$$

with a being the smoothing parameter associated with the x component and h, k are the mesh length scales in the u, v directions respectively.

Therefore, if we consider the mesh with $p = 21$ points in the u direction and $q = 21$ in the v direction, where p and q are defined as previously, then the length scales will be given by

$$\delta u = 1/20 \qquad \delta v = 2\pi/20$$

and $r^2 = a^2/(4\pi)^2$. Thus, by applying Eq. 357 at the interior mesh points and the boundary conditions at the boundary points, we obtain a linear set of equations in 19^2 unknowns for the x component of the surface. The y and z components of the surface can similarly be obtained.

Example

Consider the numerical solution of the Laplacian equation $\nabla^2 \mathbf{r} = 0$ over the rectangular domain $0 \leq u \leq 1, 0 \leq v \leq 1$, with a mesh size $1/4$ in both the u and v directions, where $\mathbf{r} = (x, y, z)$ represents the coordinates of a point in E^3. This is simply Eq. 352 with $a = 1$ and $\mathbf{f}(u, v) = 0$. Let us consider the x component of the equation when the boundary conditions are given by

$$
\begin{aligned}
x(0, v) &= \cos(\pi v) & 0 \leq v \leq 1 \\
x(1, v) &= \cos(\pi v) & 0 \leq v \leq 1 \\
x(u, 0) &= 1 & 0 \leq u \leq 1 \\
x(u, 1) &= -1 & 0 \leq u \leq 1.
\end{aligned}
$$

This defines the x component of a parametrization of a half-cylinder of unit radius as in Fig. 105a.

The finite difference scheme will be represented by

$$
\begin{pmatrix} & 1 & \\ 1 & -4 & 1 \\ & 1 & \end{pmatrix} x_{i,j} = 0 \tag{360}
$$

and when applying this equation to the mesh points illustrated in Fig. 105b we obtain the linear system of equations,

$$
\begin{aligned}
-4x_1 + x_2 + x_4 + 1 + \cos(\pi/4) &= 0 \\
x_1 - 4x_2 + x_3 + x_5 + \cos(\pi/2) &= 0 \\
x_2 - 4x_3 + x_6 + 1 + \cos(3\pi/4) &= 0 \\
x_1 - 4x_4 + x_7 + x_5 + 1 &= 0 \\
x_2 + x_4 - 4x_5 + x_6 + x_8 &= 0
\end{aligned}
$$

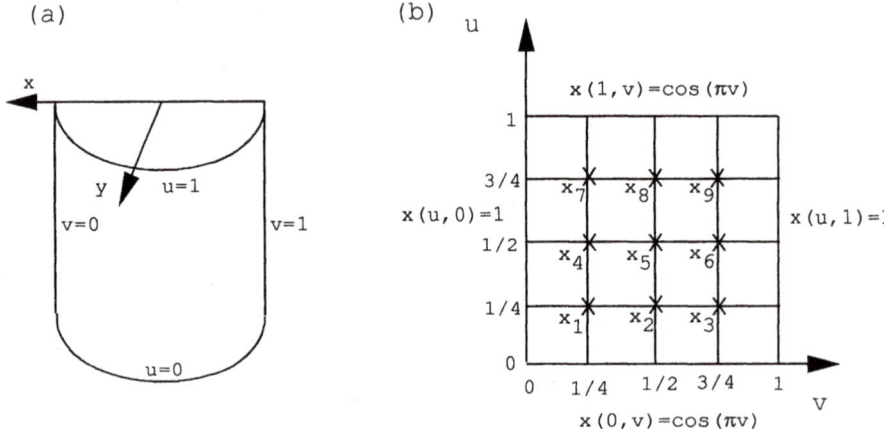

Fig. 105. The boundary conditions and grid

$$
\begin{aligned}
x_5 - 4x_6 + x_3 + x_9 + 1 &= 0 \\
x_4 - 4x_7 + x_8 + 1 + \cos(\pi/4) &= 0 \\
x_5 + x_7 - 4x_8 + x_9 + \cos(\pi/2) &= 0 \\
x_6 + x_8 - 4x_9 + 1 + \cos(3\pi/4) &= 0
\end{aligned}
$$

which can be written in the form

$$
\mathbf{A}_x \mathbf{x} = \mathbf{b}_x \tag{361}
$$

where $\mathbf{A}_x, \mathbf{b}_x$ are the matrices of coefficients associated with the solution, and $\mathbf{x}$ is a vector whose components are the values of x at the mesh points.

Similarly, suppose the y component has boundary conditions given by

$$
\begin{aligned}
y(0,v) &= \sin(\pi v) & 0 \le v \le 1 \\
y(1,v) &= \sin(\pi v) & 0 \le v \le 1 \\
y(u,0) &= 0 & 0 \le u \le 1 \\
y(u,1) &= 0 & 0 \le u \le 1
\end{aligned}
$$

and that the z component has boundary conditions

$$
\begin{aligned}
z(0,v) &= 0 & 0 \le v \le 1 \\
z(u,0) &= u & 0 \le u \le 1 \\
z(1,v) &= 1 & 0 \le v \le 1 \\
z(u,1) &= u & 0 \le u \le 1
\end{aligned}
$$

which yield two more linear systems of equations given by

$$\mathbf{A}_y \mathbf{y} = \mathbf{b}_y \qquad \mathbf{A}_z \mathbf{z} = \mathbf{b}_z. \tag{362}$$

These three systems of equations can be solved by the iterative methods as described later to produce the solution shown in Fig. 106.

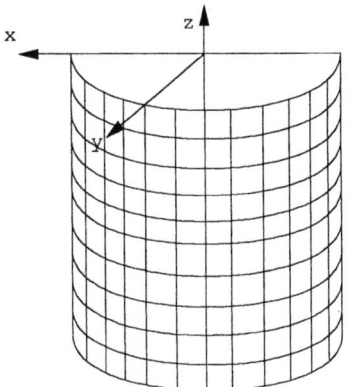

Fig. 106. The resultant surface produced by a finite difference scheme

11.3. The Fourth-Order PDE Problem

The problem illustrated in the previous example would be suitable if there were no tangency requirements at the edges of the surface. However, since there are generally requirements of tangent continuity in the cases of surface blends and free-form surfaces we require a fourth order PDE. The finite difference approximation to the 4th order operator is now given by

$$\begin{bmatrix} & & r^4 & & \\ & r^2 & -2r^2(1+r^2) & r^2 & \\ 1 & -4(1+r^2) & 6+8r^2+6r^4 & -4(1+r^2) & 1 \\ & r^2 & -2r^2(1+r^2) & r^2 & \\ & & r^4 & & \end{bmatrix} \mathbf{r} = 0 \tag{363}$$

where r is given by Eq. 359, and $f(u,v) = 0$.

If we consider this finite difference scheme applied at the mesh location $\mathbf{P}_{1,j}$ as in Fig. 107 we encounter a problem. To obtain the function value at $\mathbf{P}_{1,j}$ we need the values at the mesh points $\mathbf{P}_{2,j}, \mathbf{P}_{3,j}, \mathbf{P}_{0,j}, \mathbf{P}_{-1,j}$, etc. At the point $\mathbf{P}_{0,j}$ the function value is simply the positional boundary condition given on $\delta\Omega$. However, the point $\mathbf{P}_{-1,j}$ is external to the finite difference mesh and to obtain the corresponding function value we must consider the tangency conditions on the boundary.

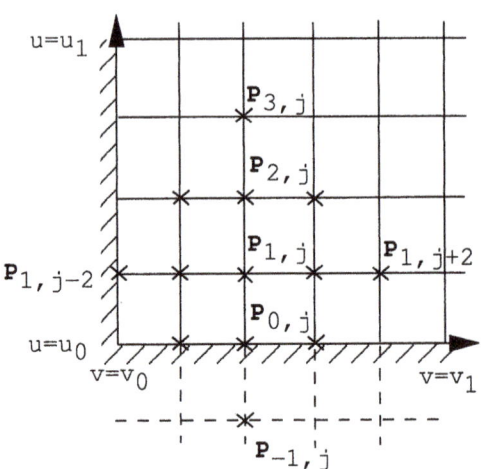

Fig. 107. Evaluation of external points by tangency conditions

11.4. Tangency Conditions

From Eq. 349 the first derivative approximation to the function x at the mesh point $\mathbf{P}_{i,j}$ will be given by

$$\left(\frac{\partial x}{\partial u}\right)_{i,j} \approx \frac{x_{i+1,j} - x_{i-1,j}}{2h}. \tag{364}$$

Now, if we consider the tangency conditions on the boundary $u = u_0$ to be given by

$$\frac{\partial x}{\partial u} = g(v), \tag{365}$$

then using the central difference formula (Eq. 364) to represent this we obtain:

$$\frac{x_{1,j} - x_{-1,j}}{2h} = g_j \tag{366}$$

where $x_{-1,j}$ is the external value at the external mesh point $\mathbf{P}_{-1,j}$. By rearranging this expression we obtain:

$$x_{-1,j} = x_{1,j} - 2hg_j. \tag{367}$$

Thus, the expression for the external mesh point can be combined with the rest of the finite difference scheme so as to obtain a linear system of equations (given by Eq. 361) that can be solved using the methods outlined below.

11.5. Numerical Approximation to a Closed Surface

In a large number of cases of blend or free-form surface design, a single patch is used to define the surface; in the examples of yacht and propeller design. In such cases the boundary conditions are often defined solely on the trimlines $u = 0$ and $u = 1$, as the solution is periodic in v.

In this case, due to the periodic boundary conditions, it can be seen from Fig. 108 that at a point on the boundary given by $v = v_0$ the solution will be the same as at corresponding points on the trimline $v = v_1$. Therefore, the finite difference scheme needs to be applied across the trimlines $v = v_0$ and $v = v_1$ as there are no explicit equations defined here.

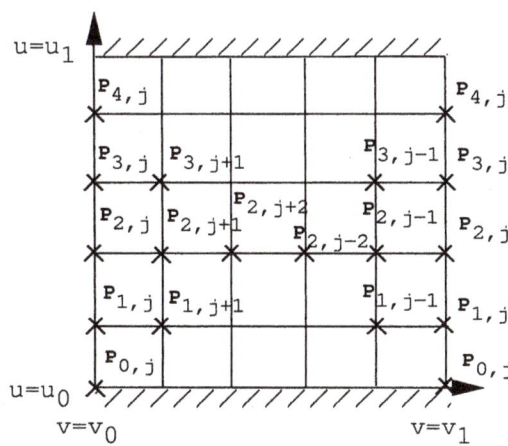

Fig. 108. Finite difference scheme at a closed boundary

If we consider the mesh point $\mathbf{P}_{2,j}$ in Fig. 108 then it can be seen that since $\mathbf{P}_{2,j} = \mathbf{P}_{2,j+q-1}$, where q is the number of mesh points in the v direction, then the 13 point finite difference scheme will be given in terms of the 13 points illustrated in Fig. 108. Note that there are no requirements for external mesh points on this boundary, as there are no derivative conditions on the $v = v_0$ and $v = v_1$.

Example

Consider the solution of $\nabla^4 \mathbf{r} = 0$ over the uniform domain $0 \leq u \leq 1$, $0 \leq v \leq 1$ with a mesh size $\frac{1}{4}$, and with boundary conditions defined by

$$
\begin{aligned}
x(0,v) &= \cos(2\pi v) & x_u(0,v) &= 0 \\
y(0,v) &= \sin(2\pi v) & y_u(0,v) &= 0 \\
z(0,v) &= h & z_u(0,v) &= S_{top}
\end{aligned}
$$

on the upper trimline, and defined on the lower trimline by

$$
\begin{aligned}
x(1,v) &= R\cos(2\pi v) & x_u(1,v) &= S_{bot}\cos(2\pi v)\\
y(1,v) &= R\sin(2\pi v) & y_u(1,v) &= S_{bot}\sin(2\pi v)\\
z(1,v) &= 0 & z_u(1,v) &= 0.
\end{aligned}
$$

These conditions produce a blend between a flat plane at a height $z = 0$ and a cylinder of unit radius centred on the z-axis with the upper trimline at a height $z = h$.

Discretization of the PDE gives the finite difference scheme, shown in Eq. 363; with $r = 1$ we obtain

$$
\begin{bmatrix}
 & & 1 & & \\
 & 1 & -4 & 1 & \\
1 & -8 & 20 & -8 & 1 \\
 & 1 & -4 & 1 & \\
 & & 1 & &
\end{bmatrix} r = 0. \tag{368}
$$

Considering now the x-component of the surface at the mesh points illustrated in Fig. 109:

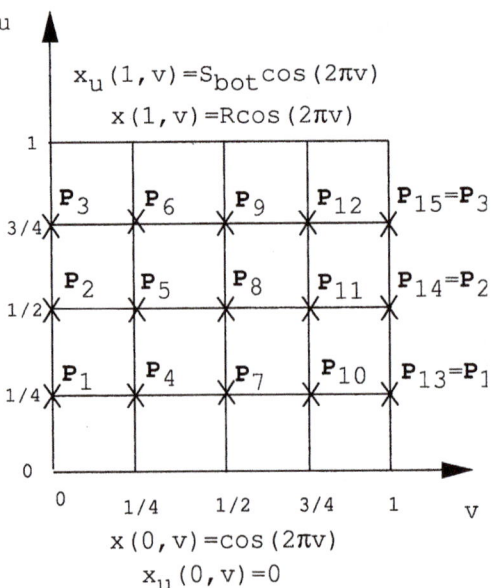

Fig. 109. The discretized domain Ω

we obtain the equation

$$
\begin{aligned}
20x_1 - 8x_2 - 8\cos(0) - 4x_4 - 4x_{10} + x_5 + x_{11}&\\
+ \cos(2\pi.3/4) + \cos(2\pi.1/4) + x_7 + x_7 + x_3 + x_{-1} &= 0
\end{aligned} \tag{369}
$$

for the mesh point P_1, with the tangency condition on the boundary $u = 0$ being defined by

$$\frac{x_1 - x_{-1}}{2.1/4} = 0 \tag{370}$$

which implies that

$$x_{-1} = x_1 \tag{371}$$

which can be substituted back into Eq. 369.

Similar equations can be derived at the other mesh points to form a linear system of equations which can be solved to give the x-component of the surface; the same procedure can be completed to derive both the y and z components of the surface blend as illustrated in Chapter 6.

12. Solution of the Linear System of Equations

For linear boundary value problems the simultaneous equations to be solved will always be linear, but the number of equations will generally be large, thus presenting quite a problem to solve them rapidly. Methods of solution include direct methods and the class of iterative methods.

Direct methods solve the system of equations in a known number of operations, with errors arising from rounding errors obtained during the computation. These methods include those of Gaussian elimination and lower-upper decomposition where the solution is sought by forward and backward substitution of a triangularly decomposed matrix [65]. However the matrix A associated with the linear equations is generally sparse and banded. This implies that there is a greater number of zero elements than non-zero elements, and that the non-zero elements lie in a band parallel to the main diagonal. Thus, iterative techniques can be more efficiently used to solve the system of equations since direct methods fill in the zero elements of the matrix with numbers used at a later stage of the process to compute the solution, whereas iterative methods ignore all elements which are zero. It is for this reason that the solution of the linear equations will be dealt with by considering iterative methods.

13. Iterative Methods

Iterative methods use an initial approximation to obtain a second, more accurate approximation to the solution, which is then in turn used, in an entirely similar fashion, to obtain a third approximation; and so on until the current approximation differs little from the previous approximate [58].

13.1. Jacobi Method

The simplest iterative method is the Jacobi method which proceeds as follows: As a simple example, consider then the set of linear simultaneous equations given by

$$a_{11}x_1 + a_{12}x_2 + a_{13}x_3 = b_1$$
$$a_{21}x_1 + a_{22}x_2 + a_{23}x_3 = b_2$$
$$a_{31}x_1 + a_{32}x_2 + a_{33}x_3 = b_3. \tag{372}$$

We can rewrite these as

$$x_1 = \frac{1}{a_{11}}(b_1 - a_{12}x_2 - a_{13}x_3)$$
$$x_2 = \frac{1}{a_{22}}(b_2 - a_{21}x_1 - a_{23}x_3)$$
$$x_3 = \frac{1}{a_{33}}(b_3 - a_{31}x_1 - a_{32}x_2) \tag{373}$$

assuming that $a_{ii} \neq 0$.

Thus, if the nth approximation to x_i is given by $x_i^{(n)}$, then the $x_i^{(n+1)}$ approximation can be found from

$$x_1^{(n+1)} = \frac{1}{a_{11}}(b_1 - a_{12}x_2^{(n)} - a_{13}x_3^{(n)})$$
$$x_2^{(n+1)} = \frac{1}{a_{22}}(b_2 - a_{21}x_1^{(n)} - a_{23}x_3^{(n)})$$
$$x_3^{(n+1)} = \frac{1}{a_{33}}(b_3 - a_{31}x_1^{(n)} - a_{32}x_2^{(n)}) \tag{374}$$

or in the general case of solving the m simultaneous equations of $\mathbf{A}x = \mathbf{b}$

$$x_i^{(n+1)} = \frac{1}{a_{ii}}\left\{ b_i - \sum_{j=1}^{i-1} a_{ij}x_j^{(n)} - \sum_{j=i+1}^{m} a_{ij}x_j^{(n)} \right\} \quad i = 1,\cdots,m. \tag{375}$$

By going through each of the m equations a new value at each of the mesh points $x_i^{(n+1)}$ can be obtained. However, the process converges only slowly [58].

13.2. Gauss-Seidel Method

As an improvement to this rather slow scheme, newly obtained values can be used in the algorithm as soon as they are determined to update the value at a neighbouring point. Therefore, if we consider Eq. 374, performed in sequence, this can now be written in the form:

$$x_1^{(n+1)} = \frac{1}{a_{11}}(b_1 - a_{12}x_2^{(n)} - a_{13}x_3^{(n)})$$
$$x_2^{(n+1)} = \frac{1}{a_{22}}(b_2 - a_{21}x_1^{(n+1)} - a_{23}x_3^{(n)})$$
$$x_3^{(n+1)} = \frac{1}{a_{33}}(b_3 - a_{31}x_1^{(n+1)} - a_{32}x_2^{(n+1)}) \tag{376}$$

and so each unknown will be a better approximation to the solution at the end of one pass through all the mesh points than at the end of a Jacobi iteration. The general case for m equations will be given by

$$x_i^{(n+1)} = \frac{1}{a_{ii}} \left\{ b_i - \sum_{j=1}^{i-1} a_{ij} x_j^{(n+1)} - \sum_{j=i+1}^{m} a_{ij} x_j^{(n)} \right\} \quad i = 1, \cdots, m. \tag{377}$$

13.3. Successive Over Relaxation Method

Despite their simplicity, the Jacobi and Gauss-Seidel methods can be so slow as to restrict their usefulness. For example, for the 5 point finite difference equation (Eq. 360), applied over the uniform mesh at 45 mesh points, the error at the end of each iteration of the Gauss-Seidel method will decrease by only a factor of around 0.995. The Jacobi method is approximately twice as slow and becomes even slower as the number of mesh points increases [58].

The method of Successive Over Relaxation is an improvement on the rate of convergence of the Gauss-Seidel method. By rewriting the Gauss-Seidel equations in the form

$$x_1^{(n+1)} = x_1^{(n)} + \left[\frac{1}{a_{11}} (b_1 - a_{11} x_1^{(n)} - a_{12} x_2^{(n)} - a_{13} x_3^{(n)}) \right]$$

$$x_2^{(n+1)} = x_2^{(n)} + \left[\frac{1}{a_{22}} (b_2 - a_{21} x_1^{(n+1)} - a_{22} x_2^{(n)} - a_{23} x_3^{(n)}) \right]$$

$$x_3^{(n+1)} = x_3^{(n)} + \left[\frac{1}{a_{33}} (b_3 - a_{31} x_1^{(n+1)} - a_{32} x_2^{(n+1)} - a_{33} x_3^{(n)}) \right] \tag{378}$$

we can see that the bracketted equations are the corrections made to $x_i^{(n)}$ by one Gauss iteration. Convergence can therefore be accelerated by giving a larger correction term to each of these components. This leads to the Successive Over Relaxation (SOR) iteration which is defined by

$$x_i^{(n+1)} = x_i^{(n)} + \frac{\omega}{a_{ii}} \left\{ b_i - \sum_{j=1}^{i-1} a_{ij} x_j^{(n+1)} - \sum_{j=i}^{m} a_{ij} x_j^{(n)} \right\} \quad i = 1, \cdots, m \tag{379}$$

or in the alternate form:

$$x_i^{(n+1)} = \frac{\omega}{a_{ii}} \left\{ b_i - \sum_{j=1}^{i-1} a_{ij} x_j^{(n+1)} - \sum_{j=i}^{m} a_{ij} x_j^{(n)} \right\} - (\omega - 1) x_i^{(n)} \tag{380}$$

$$= \omega(\text{RHS of the Gauss} - \text{Seidel iteration equations}) - (\omega - 1) x_i^{(n)}. \tag{381}$$

The factor ω is called the acceleration parameter or relaxation factor and lies in the range given by

$$1 < \omega < 2 \tag{382}$$

with $\omega = 1$ giving the Gauss-Seidel iteration equations.

For the example with 45 mesh points it can be shown [70] that the error in the SOR method will decrease by a factor of 0.87 in each iteration, which is about thirty times as fast as the Gauss-Seidel method. This becomes markedly better as the number of mesh points increases.

13.4. Convergence of Solution

One measure of the extent to which the approximate solution can be said to have converged to the actual solution, is the residual of the current approximation. The residual is a measure of the amount by which the current approximation fails to satisfy the difference equations. It is obtained by substituting the iterated values $x_i^{(n)}$ back into the original difference equation (Eq. 357) for the PDE to obtain a value R_j at each of the $j = 1, \cdots, m$ mesh points, where

$$\mathbf{R}^{(n)} \sim \mathbf{b} - \mathbf{A}\mathbf{x}^{(n)}. \tag{383}$$

The root-mean-square (RMS) value is then taken as a measure of the accuracy of the approximation of the solution to the equation. This is taken as

$$RMS = \sqrt{\frac{\sum_{j=1}^{m} R_j^2}{m}}. \tag{384}$$

When this RMS value has fallen below a given tolerance, then the solution is said to have converged.

13.5. Conjugate Gradient Method

To end this discussion of iterative processes we will look at the conjugate gradient method which is based upon a minimization technique. This method is actually a direct method in that the solution of the linear system is obtained in a finite number of steps; however, in practise, it is used as an iterative process.

To begin, note that the functional

$$\frac{1}{2}(\mathbf{x}^T \mathbf{A}\mathbf{x}) - \mathbf{x}^T \mathbf{b} \tag{385}$$

has a minimum, the solution of which is also the exact solution of the linear system $\mathbf{A}\mathbf{x} = \mathbf{b}$, where $\mathbf{x}^T$ is the transpose of the matrix $\mathbf{x}$. Then, to obtain the minimum of Eq. 385 a minimization search [65] is taken along a set of direction vectors $\mathbf{p}^k$. If we let $(\mathbf{x}, \mathbf{y}) \sim \mathbf{x}^T \mathbf{y}$ denote the Euclidean inner product, then the direction vectors $\mathbf{p}^k$ are computed so that they satisfy the condition

$$(\mathbf{p}^i, \mathbf{A}\mathbf{p}^j) = 0 \quad i \neq j, \quad i, j = 0, 1, \cdots, n-1 \tag{386}$$

and the vectors $\mathbf{p}^k$ are said to be conjugate with respect to $\mathbf{A}$.

This gives us the following conjugate gradient algorithm whose object is to obtain the minimum of the functional (Eq. 385), which gives the solution to $\mathbf{A}\mathbf{x} = \mathbf{b}$;

- Stage 1

  1. Choose a starting approximation $\mathbf{x}^0$
  2. Compute the residual $\mathbf{R}^0 = \mathbf{b} - \mathbf{Ax}^0$
  3. Set the first direction vectors to the residual $\mathbf{p}^0 = \mathbf{R}^0$ and let $k = 0$.

- Stage 2

  1. Compute $\alpha_k = (\mathbf{R}^k, \mathbf{p}^k)/(\mathbf{p}^k, \mathbf{Ap}^k)$
  2. Compute $\mathbf{x}^{k+1} = \mathbf{x}^k + \alpha_k \mathbf{p}^k$
  3. Compute $\mathbf{R}^{k+1} = \mathbf{R}^k - \alpha_k \mathbf{Ap}^k$
  4. Test for convergence. Is $\| \mathbf{R}^{k+1} \| \leq \epsilon$? If converged stop, if not go on to stage 3.

- Stage 3

  1. Compute $\beta_k = (\mathbf{R}^{k+1}, \mathbf{Ap}^k)/(\mathbf{p}^k, \mathbf{Ap}^k)$
  2. Compute $\mathbf{p}^{k+1} = \mathbf{R}^{k+1} - \beta_k \mathbf{p}^k$
  3. let $k = k + 1$
  4. Return to stage 2.

Therefore, we have an alternative method to SOR which converges to the solution at a similar rate, and thus is faster than the Gauss-seidel method [70].

In this section we have given an illustration of some of the most commonly used iterative methods for solving large sparse systems of linear equations. There are a variety of other methods for obtaining the solution to the elliptic PDEs, some of which will be discussed in the next section.

14. Collocation and Weighted Residual Methods

The finite difference methods described above can be considered as a direct discretization of differential equations. The methods described now are general techniques for obtaining approximate solutions to a PDE by a finite linear combination of known functions.

If we are given the partial differential equation

$$D^m(x) = f(u) \tag{387}$$

over the domain Ω, with boundary conditions

$$B(x) = 0 \tag{388}$$

given on $\delta\Omega$, then a weighted residual method is used to obtain an approximation to the solution by taking a trial solution of the form

$$x(u) = \sum_{j=1}^{n} c_j \phi_j(u) \tag{389}$$

where the $\phi_j(u)$ are known basis functions, which often take the form of trigonometric or polynomial functions.

It should be noted that whereas the finite difference methods obtain the unknown values at discrete mesh points within the finite domain, collocation and weighted residual methods are such that once the unknown coefficients c_j are found, they are substituted back into Eq. 389 to give the approximation to the PDE at any point inside the domain.

14.1. Collocation Method

The collocation method determines the unknowns c_j in Eq. 389 by ensuring that the approximate solution $x(u)$ satisfies Eq. 387 exactly at a set of specified points $u_1, \cdots, u_L$ in the interior of the domain Ω and, further, satisfies the boundary conditions at collocation points $u_{L+1}, \cdots, u_n$ on $\delta\Omega$. The collocation points may be taken to be uniform throughout the region or non-uniform depending on the problem. These then give n linear algebraic equations in n unknowns which can be solved to find the unknowns $c_1, \cdots, c_n$, and so the approximation to the solution $x(u)$ at the n collocation points.

Example

If we consider the linear boundary value problem

$$\frac{\partial^2 x}{\partial u^2} = 0 \tag{390}$$

with $x(0) = 0$ and $x(1) = 0$ then we can seek an approximate solution to Eq. 390 of the form:

$$x(u) = \sum_{j=1}^{n} c_j \phi_j(u) \tag{391}$$

where the basis functions ϕ_j satisfy the boundary conditions $\phi_j(0) = 0 = \phi_j(1)$ for $j = 1, \cdots, n$.

One possible example of a set of basis functions is given by

$$\phi_j(u) = u^j(1-u), \quad j = 1, \cdots, n \tag{392}$$

which gives the approximate solution to Eq. 390 in the form:

$$x(u) = u(1-u)(c_1 + c_2 u + \cdots + c_n u^{n-1}) \tag{393}$$

which is a polynomial of degree $n + 1$.

This can further be extended to the solution of Eq. 284 by considering the approximate solution

$$x(u,v) = \sum_{j=1}^{n} \sum_{i=1}^{m} c_{i,j} \phi_i(u) \phi_j(v) \tag{394}$$

with basis functions given by $\phi_i(u) = u^i(1 - u)$, $\phi_j(v) = v^j(1 - v)$ which can be thought of as a surface patch representation.

14.2. Application to the Representation of PDE Surfaces by B-Splines

It is recognised that B-spline curves and surfaces are widely used throughout the ship industry for the representation of ship forms. A relevant application of the method of collocation is the approximation of a PDE surface in terms of B-splines. The method of collocation allows us to approximate the solution of the elliptic PDE (Eq. 284) by a function of the form:

$$\mathbf{r}(u,v) = \sum_{j=-2}^{m+2} \sum_{i=-2}^{n+2} \mathbf{V}_{i,j} \cdot N_{i,k}(u) N_{j,l}(v) \tag{395}$$

where the basis functions $N_{i,k}(u)$, $N_{j,l}(v)$ are a set of quintic B-spline functions. These are used as they need be at least four-times differentiable as they must satisfy the 4th order PDE. The collocation points are chosen to be equally spaced within the interval [0,1] and correspond to the B-spline knot point which implies that the $\mathbf{V}_{i,j}$ correspond to the B-spline control points.

Therefore, we have a set of B-spline basis functions, some of whose coefficient are determined from the linear equations derived from the boundary conditions and some from the requirement that the approximation satisfies the differential equation at the collocation points. This produces a system of linear equations of the form of Eq. 361 which can be solved by any of the methods described.

Thus, for the example of the generated ship hull, it is feasible to obtain the PDE surface obtained from, say the waterlines and section curves, and so an ease of exchange of data between PDE and B-spline surfaces is available, as demonstrated in [11].

14.3. The Least Square Method of Solution

If the approximate solution is given by

$$x(u) = \sum_{j=1}^{n} c_j \phi_j(u) \tag{396}$$

then a residual can be defined as follows:

$$\epsilon(x(u)) = D^m x(u) - f(u). \tag{397}$$

So, if x_0 is the exact solution, this then implies that

$$\epsilon(x_0) = 0. \tag{398}$$

Now, a procedure for finding the unknowns c_j is by introducing a set of weighting functions $\psi_j(u)$, and to minimise the following integrals of the residual, weighted by the residual functions

$$J = \int_\Omega \epsilon_j \psi_j(u) d\Omega, \qquad j = 1, \cdots, n. \tag{399}$$

The method of least squares takes the sum of the squares of the residuals over the domain Ω to give the integral

$$J = \int_\Omega \epsilon^2 d\Omega \tag{400}$$

from which the minimum of this integral J is obtained by applying a variational principle to give the equation

$$\frac{\partial J}{\partial c_j} = 2 \int_\Omega \epsilon \frac{\partial \epsilon}{\partial c_j} d\Omega = 0, \qquad j = 1, \cdots, n. \tag{401}$$

This can be evaluated at the n points over the interior and boundary of the domain to determine the unknowns c_j. It should be noted that this is effectively equivalent to taking the weighting function in Eq. 399 as

$$\psi_j(u) = \frac{\partial \epsilon}{\partial c_j}. \tag{402}$$

14.4. Galerkin's Method

The final approach to the solution of the unknown constants c_j is obtained by choosing the weighting functions $\psi_j(u)$ to be n linearly independent functions which are required to be orthogonal to the error terms over the domain Ω, i.e.

$$\int_\Omega \epsilon \psi_j(u) d\Omega = 0, \qquad j = 1, \cdots, n. \tag{403}$$

The special case of this method is Galerkin's method for which the weighting functions $\psi_j(u)$ are taken to be the basis functions of the original approximation, thus:

$$\psi_j(u) = \phi_j(u) \tag{404}$$

which again give a system of linear equations in the n unknowns c_j from which the approximate solution can be formed.

This is the method by which Brown [11] illustrates the way in which the earlier described PDE surfaces can be represented in terms of B-splines, i.e. by using the B-spline basis functions as the weighting function.

CHAPTER 8
SURFACE GENERATION

1. Problem Statement

The design of free-form surfaces for ships, automobiles, aircraft and similar products with complex shapes, developed from functional and aesthetic criteria, requires a variety of flexible surface generation methods, suitable for different applications, design procedures and scenarios. Accordingly numerous different surface generation methods have been developed and incorporated in current systems of Computer Aided Geometric Design in recent years. This section will give a classification of problem types in surface generation, an overview of solution procedures and a few examples from specific applications. Further illustrations will follow in the next section on ship surface design.

Surface generation is a process that develops a complete mathematical surface definition from information given at discrete points or from curve information or from form parameters pertinent to the surface as a whole. The principal cases of surface construction thus proceed:

- from data at discrete points directly to the surface or

- from data at points via curves and via curve meshes to the surface or

- from global form parameters of the surface (volume, centroid etc.) and some local form parameters to the surface.

Elements of these procedures may of course be combined. All processes progress from limited information describing only certain properties of the surface to a complete and unique surface definition.

A mathematical surface definition is developed from:

- Data

- A mathematical representation

- A generation process, characterized by a criterion function and constraints.

These information elements can again serve as the basis for a taxonomy of problem types in surface definition in a similar way as for curves in Chapter 4. This classification will lead up to a unifying perspective for a large class of surface generation problems based on the format of optimization with constraints.

2. Classification

Surface generation problems can be classified according to suitable combinations of data types, mathematical representations, and generation processes (as with curve generation). This section discusses several possible choices.

a. Data:

The data types from which surfaces are constructed vary in terms of their geometric and topological properties:

- Geometric data types:

    o Points, derivatives at discrete points
    o Curves, segmented curves
    o Integral form parameters (volume etc.)

- Topological information:

    This refers to the arrangement and mutual association between the given data elements. The principal cases for surface data are (Fig. 110):

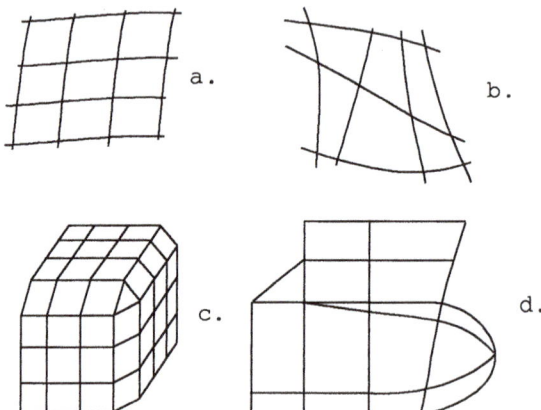

Fig. 110. Surface mesh topologies with regular mesh (a,b) and irregular mesh (c,d)

    o Regular mesh
      Data points or curves form a regular, orthogonal raster. The surface elements in this raster are bounded by four sides, they are topologically rectangular. Four edges meet at each vertex (except on the exterior boundary).
    o Irregular mesh
      Data points or curves are arranged in an arbitrary pattern so that they do not all form four-sided elements with regular corner vertices where four edges meet. The elements of the raster may be of polygonal topology, in particular triangles. An irregular data set can be triangulated so that the case of a mesh of triangles is representative of the irregular case. In some surface shapes, where singular vertices

arise, triangles cannot be completely avoided (Fig. 110, cases c. and d.)

b. Surface representations:

- Types
    - Explicit/implicit/parametric
    - Polynomial/non-polynomial
    - Integer/rational polynomials

- Requirements
    - Continuity
      Parametric continuity: $C^1, C^2, \cdots$
      Visual continuity: $VC^1, VC^2, \cdots$
      Definitions:
      VC^1 refers to continuity of the tangent plane,
      VC^2 to continuity of the principal curvatures.
    - Shape Preservation
      Positivity/monotonicity/convexity
    - Polynomial Precision
      i.e., the ability of a function to match a polynomial surface of given degrees m, n
    - Support
      Local/global
      The support of a polynomial basis function is local if the function exists only in finite intervals (of the parameters)
    - Control Mechanism
      The shape of the surface can be controlled by:
      - Offset points and derivatives
      - Control polyhedron
      - Form parameters

- Examples: Integer Parametric Polynomials
  Different polynomial bases:
    - Monomial, natural polynomial basis

$$\mathbf{r}(u,v) = \sum_{j=0}^{m} \sum_{i=0}^{n} \mathbf{a}_{i,j} u^i v^j \qquad (405)$$

    - Bézier-Bernstein basis

$$\mathbf{r}(u,v) = \sum_{j=0}^{m} \sum_{i=0}^{n} \mathbf{V}_{i,j} \cdot B_{i,n}(u) B_{j,m}(v) \qquad (406)$$

o B-Spline-Schoenberg basis

$$r(u,v) = \sum_{j=0}^{m} \sum_{i=0}^{n} \mathbf{V}_{i,j} \cdot N_{i,k}(u) N_{j,l}(v) \qquad (407)$$

- Rational Parametric Polynomials:

o NURBS basis

$$r(u,v) = \sum_{j=0}^{m} \sum_{i=0}^{n} \mathbf{V}_{i,j} \cdot R_{i,k;j,l}(u,v) \qquad (408)$$

Other bases are discussed in Section 3.

c. Generation processes

A generation process is defined by a criterion function and by constraints:

- Criteria:

A criterion sometimes used in surface fairing, in a certain analogy to the L_2 criterion for curves, is the elastic strain energy functional for a rectangular thin plate (Kirchhoff), where $f(x, z) =$ plate deflection:

$$U = \text{const} \int \int_R \left[(f_{xx} + f_{yy})^2 - 2(1-\nu) \left(f_{xx} f_{yy} - f_{xy}^2 \right) \right] dx dy \qquad (409)$$

where ν is the Poisson ratio.

This expression can be linearized and approximated by:

$$L_2 = \int \int_R \left(\kappa_1^2 + \kappa_2^2 \right) dx dy \qquad (410)$$

where κ_1, κ_2 are the principal normal curvatures as defined in Chapter 10. Similar expressions for second order criteria can also be written in parametric form [36].

- Constraints:

o Distance constraints

The sum of squares of the Euclidean distances between given discrete data points $\mathbf{P}_{i,j}$ and their associated knots $r(u_i, v_j)$ is bounded by a tolerance ϵ:

$$A = \sum_{j=0}^{m} \sum_{i=0}^{n} \left\{ w_{ij} \left(r(u_i, v_j) - \mathbf{P}_{i,j} \right) \right\}^2 - \epsilon \le 0 \qquad (411)$$

o End constraints

Applied to surface boundaries, continuously across boundary, and/or to boundary curves.

o Integral form parameter constraints

e.g., the volume under a surface or the volume centroid may be assigned given values.

Examples

Surface generation processes are for example:

- Approximation of scattered data:
    Data: Discrete points, irregular mesh
    Representation: e.g. triangular surface patches
    Process: Fairness norm, distance constraint

- Interpolation of irregular data:
    As above, however distance constraint equal to zero: Discrete data interpolation.

- Transfinite, regular curve mesh interpolation:
    Data: regular mesh of curves
    Representation: e.g. Coons Boolean sum parametric polynomial patches
    Process: Equality distance constraint ($= 0$) on mesh lines. Possibly fairness norm.

More examples are given in Section 3. Fig. 111 gives an overview of surface representations and corresponding interpolatory generation processes. Transfinite interpolation refers to the exact interpolation of the given curve mesh; discrete interpolation accounts for given data only at discrete points.

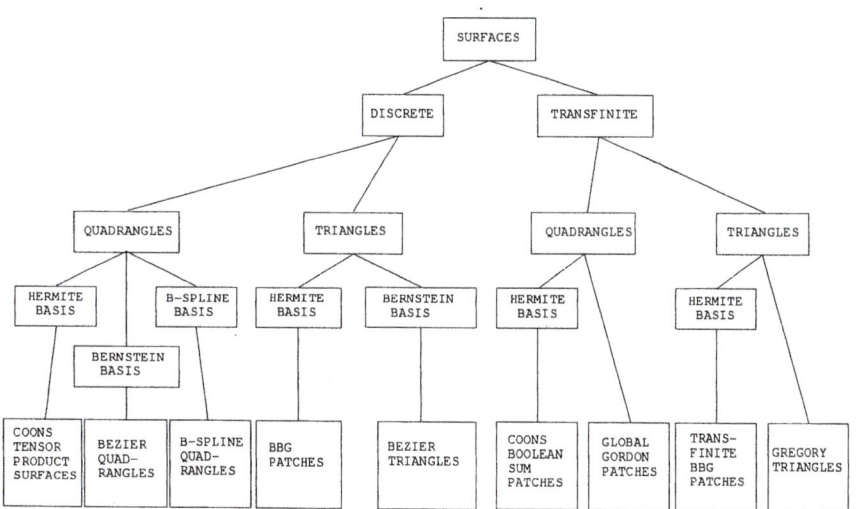

Fig. 111. Classification of surface interpolation for discrete and transfinite cases (*BBG: Barnhill-Birkhoff-Gordon)

3. Surface Representations

3.1. Coons Boolean Sum Interpolation

The Boolean sum surface construction by S. A. Coons [18] is a transfinite interpolation method for a single quadrilateral patch (Fig. 112), hence a local interpolant of four given boundary curves. The patch has curvilinear coordinates u and v. Coons uses the abbreviated notation:

$\mathbf{r}(u, v) = (u \ v)$ = surface equation

$\mathbf{r}(u, 0) = (u \ 0)$ = boundary curve,$v = 0$

$\mathbf{r}(1, v) = (1 \ v)$ = boundary curve,$u = 1$

$\mathbf{r}(0, 0) = (0 \ 0)$ = corner point,$u = v = 0$

etc.

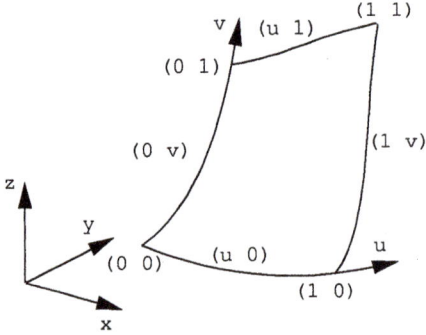

Fig. 112. Coons patch

Coons develops a Boolean sum expression by equating the surface to the Boolean sum set which contains all four patch boundary curves.

If $A(F)$ and $B(F)$ are two set operators acting on F, then the Boolean sum $(A \oplus B)(F)$ is defined as the union of the two sets (Fig. 113), which contains the intersection set $A(B(F)) = B(A(F)) = AB(F)$ only once. Hence

$$(A \oplus B)(F) = A(F) + B(F) - AB(F). \tag{412}$$

Coons has applied this operator in a geometric context such that

$$A = P_u[\mathbf{r}(u, v)] = \mathbf{r}(0, v)F_0(u) + \mathbf{r}(1, v)F_1(u) \tag{413}$$

A(F),B(F)=two sets

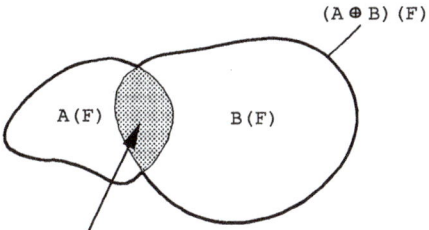

(A ⊕ B) (F)

A(F)   B(F)

Intersection =A(B(F))=A.B(F)=B.A(F)
set

Fig. 113. Boolean sum of two sets

which equals the interpolant in the $u-$direction of the boundaries at $u = 0$ and $u = 1$, where $F_0(u), F_1(u)$ are Hermite interpolants to the boundaries at $u = 0$ and $u = 1$. The degree of these Hermite polynomials is chosen to suit the boundary conditions. Similarly for $F_0(v), F_1(v)$. Coons calls these polynomials blending functions.

$$B = P_v[\mathbf{r}(u,v)] = \mathbf{r}(u,0)F_0(v) + \mathbf{r}(u,1)F_1(v) \tag{414}$$

is the interpolant in the $v-$direction of boundaries at $v = 0$ and $v = 1$.

$$\begin{aligned} AB &= P_u\left[P_v[\mathbf{r}(u,v)]\right] \\ &= \mathbf{r}(0,0)F_0(u)F_0(v) + \mathbf{r}(0,1)F_0(u)F_1(v) \\ &+ \mathbf{r}(1,0)F_1(u)F_0(v) + \mathbf{r}(1,1)F_1(u)F_1(v). \end{aligned} \tag{415}$$

Substituting we get

$$\begin{aligned} \mathbf{r}(u,v) &= (uv) = (P_u \oplus P_v)[(uv)] \\ &= (0v)F_0(u) + (1v)F_1(u) + (u0)F_0(v) + (u1)F_1(v) - (00)F_0(u)F_0(v) \\ &- (01)F_0(u)F_1(v) - (10)F_1(u)F_0(v) - (11)F_1(u)F_1(v). \end{aligned} \tag{416}$$

Condensed into matrix notation this Boolean sum patch becomes

$$(uv) = -\begin{bmatrix} -1 & F_0(u) & F_1(u) \end{bmatrix} \begin{bmatrix} 0 & u0 & u1 \\ 0v & 00 & 01 \\ 1v & 10 & 11 \end{bmatrix} \begin{bmatrix} -1 \\ F_0(v) \\ F_1(v) \end{bmatrix} \tag{417}$$

where $ij = (i\ j) = v(i,j)$, taken for $u = i, v = j$.

This type of patch interpolates the given boundary curves exactly regardless of their representation. The blending functions are chosen of sufficient degree in such a way that a change made at one boundary does not affect the opposite boundary of the patch. This requires:

$$F_0(0) = 1, F_0(1) = 0, F_1(0) = 0, F_1(1) = 0 \tag{418}$$

and to keep the first derivatives at a boundary independent of the shape of the opposite boundary, too:

$$F_0'(0) = F_1'(0) = F_0'(1) = F_1'(1) = 0. \tag{419}$$

For these constraints cubic blending functions $F_0(u), F_1(u), F_0(v), F_1(v)$ will do.

This Boolean sum patch complies with given boundary curves, but does not account for derivatives along the boundaries. If these are specified, too, an extended Coons Boolean sum patch can be constructed. For example, if the first derivatives across the boundary curves and the mixed partial derivatives at the corner points are given, then the form of the Boolean sum patch is:

$$(uv) = -[\ -1 \quad F_0(u) \quad F_1(u) \quad G_0(u) \quad G_1(u)\ ] \cdot B \cdot \begin{bmatrix} -1 \\ F_0(v) \\ F_1(v) \\ G_0(v) \\ G_1(v) \end{bmatrix} \tag{420}$$

with

$$B = \begin{bmatrix} 0 & u0 & u1 & u0_v & u1_v \\ 0v & 00 & 01 & & \\ 1v & 10 & 11 & & \\ 0v_u & & & 00_{uv} & 01_{uv} \\ 1v_u & & & 10_{uv} & 11_{uv} \end{bmatrix}. \tag{421}$$

The functions G_0, G_1 are further blending functions subject to

$$G_0(0) = G_0(1) = G_1(0) = G_1(1) = 0$$
$$G_0'(0) = 1; G_0'(1) = 0; G_1'(0) = 0; G_1'(1) = 0. \tag{422}$$

This kind of Boolean sum patch interpolates to the patch boundaries and to the given cross-boundary tangent directions.

3.2. Coons Cartesian Product Surface

In the special case that the boundary curves can be completely defined by information about position and derivative properties at their ends, i.e., at the corner points of the patch, the Boolean sum representation can be transformed into Cartesian product form, also called tensor product form, which interpolates only the discrete data at the patch corners. Discrete interpolation of a quadrilateral patch is thus developed as a special case of transfinite interpolation.

The result of this development is the Coons Cartesian product patch, which is of the form:

$$(uv) = [\ F_0(u) \quad F_1(u) \quad G_0(u) \quad G_1(u)\ ] \cdot B_C \cdot \begin{bmatrix} F_0(v) \\ F_1(v) \\ G_0(v) \\ G_1(v) \end{bmatrix} \tag{423}$$

with

$$
B_C = \begin{bmatrix} 00 & 01 & 00_v & 01_v \\ 10 & 11 & 10_v & 11_v \\ 00_u & 01_u & 00_{uv} & 01_{uv} \\ 10_u & 11_u & 10_{uv} & 11_{uv} \end{bmatrix}.
\tag{424}
$$

This patch requires and interpolates positional and first derivative data (boundary curve tangents) at the corner points as well as mixed derivatives (called twist vectors) also at these points, e.g:

$$
00_{uv} = \frac{\partial^2 \mathbf{r}(u,v)}{\partial u \partial v} \text{ at } u = 0, v = 0.
$$

The blending functions are chosen from the same considerations as before. If conditions 418, 419 and 422 are applied, then cubic blending functions result. This yields the frequently used Coons bicubic tensor product patch, in which the functions $F_0, \cdots, G_1$ are a set of cubic Hermite interpolants.

To construct such a patch the geometry matrix B_C must be filled with data. The corner point positions and the first partial derivatives at the corner points can be deduced from the boundary curves, and evaluated at these discrete points. The numerical determination of twist vector values at patch corners is more difficult because this quantity is a surface property, not a boundary curve attribute. An approximation to this value can be derived from drawing data, where the Coons patch may be part of a curve mesh, by numerically differentiating (finite difference quotient) the first partial derivative vector, e.g. $(0v)_u$, with respect to the other derivative direction. This process is inaccurate, but still much better than just equating the twist vector partition to zero as is sometimes done. In some cases recently the twist vectors have also been determined from optimum surface fairness requirements (Kallay and Ravani [36]).

It can be shown that the bicubic Coons patch can be joined together with other such patches in a mesh so that tangent plane continuity can be achieved.

3.3. Bézier Surfaces

A quadrangular Bézier surface [5] is defined by

$$
\mathbf{r}(u,v) = \sum_{j=0}^{m} \sum_{i=0}^{n} \mathbf{V}_{i,j} \cdot B_{i,n}(u) B_{j,m}(v)
\tag{425}
$$

where

$\mathbf{V}_{i,j}$ = controlling vertices, Bézier points, in regular mesh arrangement of $(m+1)(n+1)$ points

$B_{i,n}(u)$ = Bernstein polynomial in u of degree n (Chapter 3, Section 3.3)

$B_{j,m}(v)$ = Bernstein polynomial in v of degree m.

Bézier surfaces are a direct bivariate generalization of Bézier curves. They thus inherit many of the properties demonstrated for Bézier curves in Chapter 3, Section 3.4. This is based on the properties of the Bernstein polynomial basis and includes the following surface properties:

-   Invariance to coordinate transformations

-   Variation diminishing property

-   Convex hull property

-   End vertex interpolation

-   End tangent control by last polygon leg

-   Degree $n=$ number of vertices minus one

-   Global shape control

At the same time a Bézier surface is also a tensor product surface and hence suitable as a discrete interpolant. The surface can be required to interpolate as many discrete data (of vector type) as there are free vertices. It cannot match arbitrary given boundary curves exactly (transfinite case), but it can take into account boundary curves that are themselves Bézier curves of corresponding degree.

The difference between Coons Cartesian product and Bézier surfaces lies principally in the different polynomial basis, Hermite versus Bernstein. A given basis function can be exactly converted to the other polynomial basis of the same degree by equating coefficients. Therefore the two surface representations are analytically fully equivalent, which holds for other tensor product surfaces (like B-Spline surfaces) as well. This means that the resulting geometric shapes are also equivalent and can be matched in different tensor product representations.

The main distinction of the Bézier representation thus lies in its form of shape control by vertex manipulation in the defining polygon, which is intuitively very appealing in interactive work. This is of particular advantage when surfaces are constructed interactively on the monitor screen, but does not prevent meeting certain discrete point interpolation constraints.

3.4. B-Spline Surface

A quadrangular B-Spline surface is defined by

$$\mathbf{r}(u,v) = \sum_{j=0}^{m}\sum_{i=0}^{n} \mathbf{V}_{i,j} \cdot N_{i,k}(u)N_{j,l}(v) \tag{426}$$

where

$\mathbf{V}_{i,j}$ =controlling vertices, de Boor points, in regular mesh arrangement of $(n+1) \cdot (m+1)$, $(n+1) \geq k, (m+1) \geq l$

$N_{i,k}(u)$ =B-spline basis function in u of degree $(k-1)$ (Chapter 3, Section 3.7)

$N_{j,l}(v)$ =B-spline basis function in v of degree $(l-1)$

B-spline surfaces are a direct bivariate generalization of B-spline curves. They therefore share many of the properties demonstrated in Chapter 3, Section 3.9 for B-spline curves. These properties are based on the B-spline basis function and include the following:

- Invariance to coordinate transformation

- Variation diminishing property

- Convex hull property

- End vertex interpolation

- End tangent control by last polygon leg

- Degrees = order minus one: $(k-1),(l-1)$

- Number of de Boor points free: $n+1 \geq k, m+1 \geq l$

- Local shape control

- Equivalence to Bézier surface for $n+1 = k, m+1 = l$

B-spline surfaces are another kind of tensor product surfaces, too. They can thus serve as discrete interpolator to as many given discrete data (of vector type) as there are $(n+1)\cdot(m+1)$ free vertices. The surface has B-spline curves as boundary curves. Fig. 114 shows an example of a B-spline surface and its defining polyhedron.

B-spline surfaces may consist of several segments in both the $u-$ and $v-$directions if the number of defining vertices exceeds the order of the B-spline polynomials ($n+1 > k$ and/or $m+1 > l$). In this case internal segment boundaries or knot lines exist. The control of the surface shape is local, i.e., the displacement of some vertex does not necessarily affect all segments. The greater the surplus of vertices over the polynomial order, the more segments are created and the greater the extent of local control.

The differentiability of segment boundaries or knot lines for simple knots equals $(k-2)$ and $(l-2)$, respectively. That is, the surface is then of continuity C^{k-2} and C^{l-2} at these boundaries. For multiple knots the differentiability decreases by one for each increase in multiplicity. Multiple knots can thus be used, e.g., to generate local knuckles or curvature discontinuities in surfaces that are otherwise tangent plane and curvature continuous.

Fig. 115 shows an example for such an application. The surface has three segments separated by two simple knot lines in its lower part and is of order $k=3$, hence quadratic and tangent plane continuous in that direction there. In the upper parts

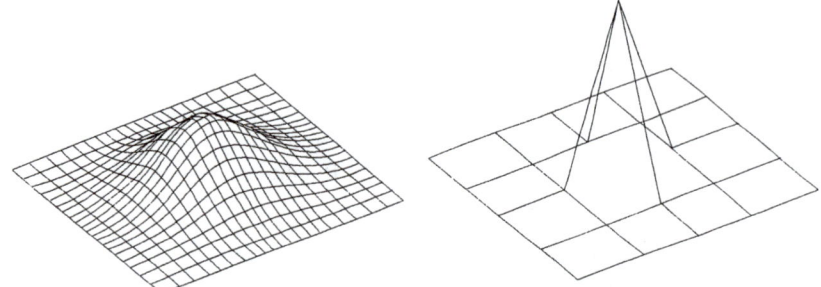

Fig. 114. B-Spline surface and defining polyhedron

the two knot lines coincide and are forming a double knot so that a knuckle (tangent plane discontinuity) is generated.

B-spline surfaces can be converted into other kinds of tensor product surfaces by equating coefficients. They are thus analytically and geometrically largely equivalent to other representations of this type. Yet their special polynomial basis gives them unique capabilities in surface generation in terms of defining polygon shape manipulation, local control and continuity control. This is why they are widely accepted in geometric modelling systems.

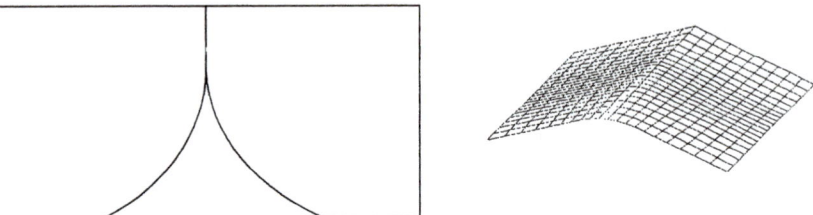

Fig. 115. B-Spline surface with double knot line

3.5. Rational B-Spline (NURBS) Surfaces

The rational B-spline surface is defined as

$$\mathbf{r}(u,v) = \frac{\sum_{j=0}^{m}\sum_{i=0}^{n}\mathbf{V}_{i,j}N_{i,k}(u)N_{j,l}(v)w_{ij}}{\sum_{j=0}^{m}\sum_{i=0}^{n}N_{i,k}(u)N_{j,l}(v)w_{ij}} = \sum_{j=0}^{m}\sum_{i=0}^{n}\mathbf{V}_{i,j}\cdot R_{i,k;j,l}(u,v) \qquad (427)$$

where

$\mathbf{V}_{i,j}=$ controlling vertices in regular mesh arrangement of $(n+1)\cdot(m+1)$, $(n+1)\geq k$, $(m+1)\geq l$

$$R_{i,k;j,l}(u,v) = \frac{N_{i,k}(u)N_{j,l}(v)w_{ij}}{\sum_{i=0}^{n}\sum_{j=0}^{m}N_{i,k}(u)N_{j,l}(v)w_{ij}} = \text{Rational B-spline function in } u,v$$

$$\text{of degree } (k-1)(l-1) \tag{428}$$

$w_{ij}=$ weight associated with control vertex $\mathbf{V}_{i,j}$.

NURBS surfaces are a direct bivariate generalization of NURBS curves. They therefore share many of the properties demonstrated in Chapter 3, Section 3.15.

4. Surface Generation Processes

4.1. Fairing Process

Surface generation processes are characterized by a criterion function and constraints. A fairing process uses a fairness measure as its criterion. It is an intriguing thought that the fairing processes based on elastic fairness measures for curves should be extended to surfaces in order to achieve 'optimal' surface shapes in a single, surface-based fairing process. In analogy to curves the fairness criterion for a surface would be the minimization of the strain energy in the surface modelled as a thin elastic plate. Physically the deformation of a plate corresponds to its equilibrium position under given constraints. Whether this shape is also advantageous and in the designer's true interest for any application is an open question and depends on the functional objectives of the design. Alternative criteria on a more functional basis, e.g., the shape of least drag for a ship surface, should therefore be explored. Meanwhile it can be held that elastic fairness measures may serve to generate surfaces of at least favourable continuity properties not unlike traditional fairing processes for curves.

According to Section 2 (Eq. 410) the elastic strain energy functional in a rectangular thin plate (Kirchhoff) can be approximated by the criterion

$$L_2 = \int\int_R \left(\kappa_1^2 + \kappa_2^2\right) dx dy \tag{429}$$

where κ_1, κ_2 are the principal normal curvatures.

This criterion, or further approximations thereof, have been used in the development of several direct elastic surface fairing processes. Walter [78] used it in connection with a mesh of bicubic Coons patches, Nowacki/Reese [56] and Hagen/Schulze [29] used bicubic Coons patches instead. Kallay/Ravani [36] went back to bicubic Coons patches and determined the optimal twist vectors.

Surface fairing is usually performed for a given mesh of boundary curves and sometimes even for given tangent vector and curvature distributions as end constraints at the exterior boundaries. The surface elements in this situation are no longer freely deformable. In fact, in Coons tensor product form only the mixed partial derivatives (twist vectors etc.), in Bézier or B-spline representation only the interior control vertices are still free to be optimized.

Further details about the results of the fairing processes are given in the references cited above. Based on our experience the following observations seem to be justified:

- The fairness criterion L_2 and hence the success of the fairing process to a great degree depend on the fairness of the mesh of boundary curves which are interpolated by the surface. Deficiencies in boundary curve fairness cannot be corrected by the surface fairing process. Therefore high priority must be placed on pre-fairing the mesh of boundary curves.

- Once the boundary curves are set, the improvements in the fairness measure (L_2) which can be achieved in surface fairing by optimizing the twist vectors or interior control vertices are usually only limited. Yet arbitrary assignments to these free variables, e.g., null twist, can result in considerable unfairness (wrinkles) in the surface. This is why some surface fairing process should not be neglected.

4.2. Other Generation Processes

One of the difficulties of surface generation is that of using two-dimensional information to create a surface in 3 dimensions. Techniques have been devised to generate surfaces that use a 2 dimensional curve following some trajectory in space. This gives an intuitive feel to the surface design [81].

o Translational Sweeping:
This is one of the most basic examples of using a curve to build a surface. The surface is generated by taking some profile, or generating curve (generatrix), and sweeping it along a trajectory, or guiding curve (directrix). The directrix may be a straight line or curved, with the scale of the generatrix able to vary as it moves.

Consider the example of generating a cylinder. This we may do by sweeping a circular profile curve along a straight line. Thus, if the profile curve is taken as the NURBS representation of a circle in Chapter 3 (Eq. 193), on the knot vector $\mathbf{V}$ with weights w_i, then we can denote the parameter for the sweep direction by u, where $0 \leq u \leq 1$. The surface of the cylinder will then be defined by:

$$\mathbf{r}(u,v) = \sum_{i=0}^{1} \sum_{j=0}^{8} \mathbf{V}_{i,j} \cdot R_{i,2;j,3}(u,v) \qquad (430)$$

where $\mathbf{U} = (0,0,1,1)$ is the knot vector in the u direction.
Note that at the ends $u = 0$ and $u = 1$ the control points will be given by the

control points of the circle, and similarly the weights will be those of the circle, i.e. $w_{0j} = w_{1j} = w_j$ where w_j are those associated with Eq. 193 (Fig. 116).

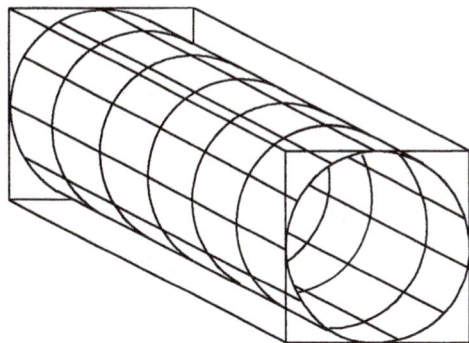

Fig. 116. Cylindrical surface

o Ruled Surface:

If we have two general NURBS curves given by

$$r_1(u) = \sum_{i=0}^{n} V_i^1 \cdot R_{i,k}(u) \quad r_2(u) = \sum_{i=0}^{n} V_i^2 \cdot R_{i,k}(u) \tag{431}$$

defined over the knot vectors U_1 and U_2, then we define a ruled surface to be a linear interpolation between $r_1(u)$ and $r_2(u)$. This produces a surface which has straight line segments connecting the points of equal parameter value. For the surface to be constructable, the two curves $r_1(u)$ and $r_2(u)$ must have the same degree $(k-1)$ and be defined over the same knot vector and parameter range. If this is not the case, then the curves can be modified using the techniques described in the next section. The surface will then be defined by:

$$r(u,v) = \sum_{i=0}^{n} \sum_{j=0}^{1} V_{i,j} \cdot R_{i,k;j,2}(u,v) \tag{432}$$

where $V = (0,0,1,1)$, and $V_{i,j}$ is the control polygon obtained from merging the control polygons of r_1 and r_2. The surface can be seen in Fig. 117.

o Surface of Revolution:

A surface of revolution may be designed by taking a profile curve,

$$r(u) = \sum_{i=0}^{n} V_i \cdot R_{i,k}(u) \tag{433}$$

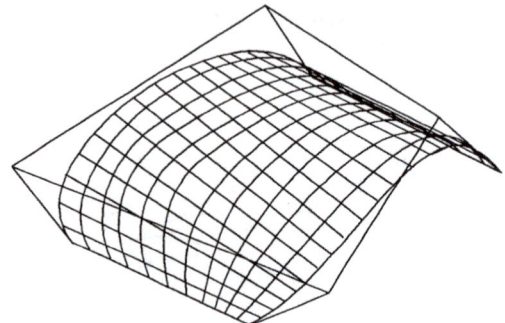

Fig. 117. A ruled surface

over the knot vector $\mathbf{U}$, lying in the (x, z) plane say, and rotating it about the z-axis to form either an open or closed surface. A full surface of revolution is generated by rotating $\mathbf{r}(u)$ by 2π radians, and so we combine the profile curve $\mathbf{r}(u)$ with the NURBS representation for a circle, thus giving a surface of the form:

$$\mathbf{r}(u, v) = \sum_{i=0}^{n} \sum_{j=0}^{8} \mathbf{V}_{i,j} \cdot R_{i,k;j,3}(u, v) \tag{434}$$

where the knot vector $\mathbf{V}$ is the circle knot vector of Eq. 193. For $i = 0$, the control points $\mathbf{V}_{i,j}$ of the surface are the control points $\mathbf{V}_j$ of the circle. The weights w_{ij} are defined by multiplying the weight values of $\mathbf{r}(u)$ with the weights of the circle, i.e. for fixed $i, w_{i0} = w_i \cdot 1, w_{i1} = w_i \cdot \sqrt{2}/2, w_{i2} = w_i \cdot 1$ etc. Fig. 118b illustrates the surface of revolution obtained from rotating the profile curve of Fig. 118a.

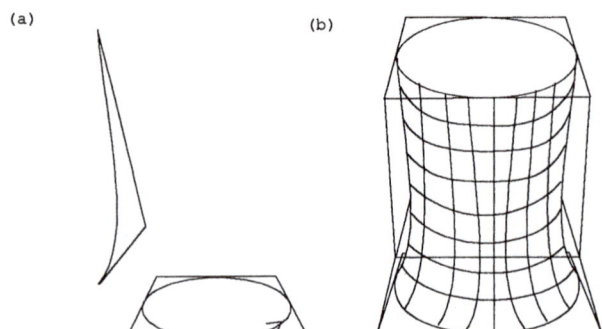

Fig. 118. A surface of revolution

o Swinging:

Swinging, or free-form sweeping is a generalization of the surface of revolution described in the last section. If we maintain a profile curve $\mathbf{r}(u)$ in the (x, z) plane say, given by Eq. 433 then instead of a trajectory around the circumference of the circle, we define some new trajectory curve:

$$\mathbf{t}(v) = \sum_{j=0}^{m} \mathbf{T}_j \cdot R_{j,l}(v) \qquad (435)$$

about which to swing $\mathbf{r}(u)$. Swinging the profile curve about the z-axis then yields the surface:

$$\mathbf{r}(u, v) = \sum_{i=0}^{n} \sum_{j=0}^{m} \mathbf{V}_{i,j} \cdot R_{i,k;j,l}(u, v) \qquad (436)$$

where the weights w_{ij} are given by the product of the weights of $\mathbf{r}(u)$ and $\mathbf{t}(v)$, i.e. $w_{ij} = w_i \cdot w_j$, and $\mathbf{V}_{i,j}$ correspond to the positions of the control points of the trajectory and profile curves (see Fig. 119).

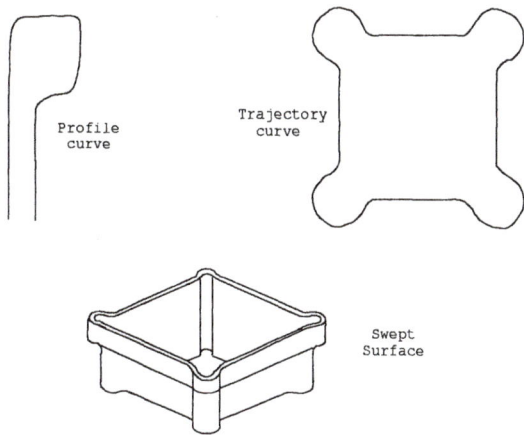

Profile curve

Trajectory curve

Swept Surface

Fig. 119. A swept surface

o Lofting:

In lofting a set of planar curves (cross sections) is given as a function of one parametric variable u, i.e., $\mathbf{r}_i = \mathbf{f}_i(u)$. They are interpolated by a lofting function $g(v)$ in the direction of the other surface variable v. The interpolation may be performed by Lagrange interpolants $g_i(v)$ such that

$$\mathbf{r}(u, v) = \sum_{i=0}^{n} \mathbf{f}_i(u) \cdot g_i(v) . \qquad (437)$$

4.3. Skinning Methods

Skinning is a special process of surface generation based on the interpolation of a given set of curves. In one sense it is very similar to lofting in that the surface here, too, interpolates a family of predefined curves. In another it is more specific by using B-spline representations for both the curves and surfaces. The given curves are frequently planar cross sections of the design objects so that they are easy to generate from a drawing, sketch or interactively on the screen. However, the skinning process can also be applied to a set of non-planar curves. Skinning results in the generation of a B-spline surface [82].

Skinning usually begins with the definition of an axis curve, called the spine or skeleton curve. It is an arbitrary, continuous space curve which serves as the reference curve for the set of given interpolation curves. Each interpolation curve can be related to a parameter value or knot in the spine. The spine is represented as a B-spline curve:

$$\cdot(t) = \sum_{i=0}^{n} \mathbf{V}_i \cdot N_{i,k}(t). \tag{438}$$

This spine curve can be derived by interpolation of some given data points for this axis. The parametrization should be non-uniform in close correspondence to arc length or chord length between knots to avoid problems with curve oscillations or unfairness.

The given set of interpolation curves describes cross sections through the shape to be generated. The curves may be planar or spatial, open or closed, and are non-intersecting with each other. They should be given in B-spline representation, too, or if they are initially given as parameter polynomials of some other basis, then they should be transformed to B-splines by appropriate basis conversion [33].

Each intersection curve has its own reference plane and a local origin and coordinate system in this plane. It is defined relative to this coordinate system. The local reference system is then attached to the spine at certain knots t_i in such a way that the reference plane is in the normal plane of the spine curve and so that the local origin either lies on the spine or is defined relative to the spine by a given translation. The local coordinate system may also be rotated within the normal plane, e.g. in the direction of the normal and binormal vectors of the spine curve or so that the axes projections are parallel to two global Cartesian coordinate axes. Let us suppose in the following that these appropriate translations and rotations have already been performed so that we refer to the intersection curves in their intended spatial position and orientation.

Let a given interpolation curve (skin curve) associated with the spine parameter value t_i be denoted by Fig. 120:

$$\mathbf{r}_i(s) = \sum_{j=0}^{m} \mathbf{V}_j^{(i)} \cdot N_{j,l}(s) \tag{439}$$

for $0 \leq i \leq n$, and where $\mathbf{V}_j^{(i)}$ are control points of skin curve i (usually determined by interpolating some given data points), and $N_{j,l}(s)$ is a B-spline basis function of degree $(l-1)$.

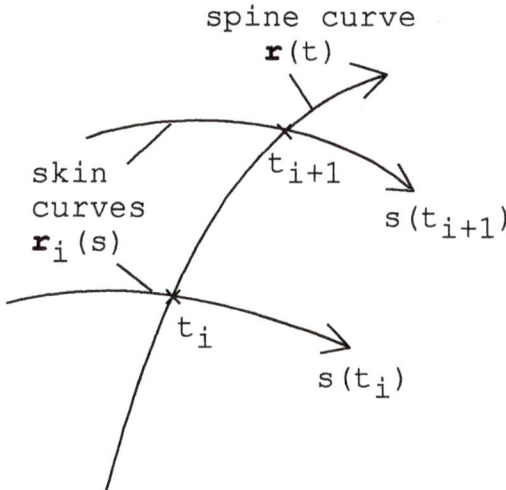

Fig. 120. Spine curve and skin curve parametrization

These skin curves are now to be interpolated in the direction of the spine axis parameter t in order to obtain a B-spline surface. This meets with a fundamental requirement: The skin curves must be of equivalent segmentation, parametrization and polynomial degrees in order to achieve the standard tensor product form of a B-spline surface. Initially, however, many skin curves differ in some of these respects. They must thus be subjected to some transformations in order to unify their segmentation, parametrization and polynomial degrees before they can be interpolated in the other parametric direction.

This is why the skinning approach goes through the following sequence of steps to prepare and perform the surface interpolation:

- Normalization of skin curves:
    All skin curves, which are represented as B-splines, Eq. 439, are normalized to the same parameter range, $s_0 \leq s \leq s_1$, and parameter orientation. This is done by linear reparametrization resulting in identical parameter ranges (s_0, s_1).

- Standardization of degrees:
    All polynomial degrees in the set of skin curves are equalized. This is done by degree elevation to the highest degree present (precise, but higher computer time and storage space) or by degree reduction to the lowest occurring degree (only approximate, subject to tolerances, but economical in time and storage space) or by choosing some degree in between. A new knot vector results for each transformed curve from these operations.

Algorithms for B-spline degree elevation and degree reduction are found in the literature [33], [62], [20], [63]. Prautzsch and Piper [63] have presented a fast algorithm for degree raising of spline curves.

After these transformations all degrees in the skin curve set are uniform.

- Homogenization of knot vectors:

  At this stage initially the knot vectors $\mathbf{T}_i$ of the skin curves may be different. They need to be brought into agreement. This is done by defining a common knot vector $\mathbf{T}$ for all curves which contains the complete set of knots of the individual curves.

  Then the individual curves are adjusted to the common knot vector by inserting any missing knots so that all curves now have the same knot vector $\mathbf{T}$. Knot insertion is achieved by algorithms derived by Boehm [9] or by Cohen et al. [19], the Oslo algorithm (See Chapter 3, Section 3.11). This process does not alter the curve shape.

  Knot insertion may increase the data volume significantly. For example, if all $m + 1$ knots in each of $n + 1$ curves were different initially (worst case), then the common knot vector $\mathbf{T}$ would have in the order of $(m + 1)(n + 1)$ knots, hence for all curves together $(m + 1)(n + 1)^2$ knots. This may result in prohibitive data volumes. It is advisable therefore to parametrize the knots in the curve set as uniformly as possible.

  In addition, if the data volume is still too large, it is possible to apply knot removal approximations [42], [33] within given tolerances to economise on the data. This process must be applied uniformly to all skin curves.

  In the end of this step all skin curves are represented with uniform knot vector, parametrization, and degree. Their form is still of the representation type in Eq. 439, though with a new standardized knot vector and corresponding control points.

- Skinning interpolation:

  The set of skin curves thus prepared now possesses an identical number $m + 1$ of correlated control points $\mathbf{V}_j^{(i)}$ for each curve. These control points are now interpolated with respect to the parameter t by fitting a set of 'longitudinal curves' $\mathbf{r}_j(t)$ through the control points, one for each j:

$$\mathbf{r}_j(t) = \sum_{i=0}^{n} \mathbf{V}_i^{(j)} \cdot N_{i,k}(t) \tag{440}$$

where $\mathbf{V}_i^{(j)}$ are control points of the longitudinal curve j, $N_{i,k}(t)$ is a B-spline basis function of order k, such that the skin curves are interpolated at their knot parameters t_i:

$$\mathbf{r}_j(t_i) = \sum_{i=0}^{n} \mathbf{V}_i^{(j)} \cdot N_{i,k}(t_i) = \mathbf{V}_j^{(i)} \tag{441}$$

for $0 \leq j \leq m$.

Combining Eq. 439 and Eq. 440 into a B-spline surface equation yields the form:

$$\mathbf{r}(s,t) = \sum_{i=0}^{n} \sum_{j=0}^{m} \mathbf{V}_i^{(j)} \cdot N_{i,k}(t) N_{j,l}(s). \qquad (442)$$

That is, if the control points $\mathbf{V}_i^{(j)}$ from the longitudinal pass in Eq. 440 are used in Eq. 442 to define the control mesh for the skinning surface, then the skin curves are exactly interpolated, as resubstitution of $t = t_i$ in Eq. 442 demonstrates:

$$
\begin{aligned}
\mathbf{r}(s,t_i) &= \sum_{i=0}^{n} \sum_{j=0}^{m} \mathbf{V}_i^{(j)} \cdot N_{i,k}(t_i) N_{j,l}(s) \\
&= \sum_{j=0}^{m} \mathbf{r}_j(t_i) \cdot N_{j,l}(s) \\
&= \sum_{j=0}^{m} \mathbf{V}_j^{(i)} \cdot N_{j,l}(s) = \mathbf{r}_i(s).
\end{aligned}
\qquad (443)
$$

Example: Propeller Geometry

The geometry of a marine propeller blade can be readily represented by a B-spline surface using the skinning approach. In this case the spine curve corresponds to the generating line of the propeller blade (with rake, skew etc.). The skin curves are taken as the foil sections, i.e., cylindrical section (constant radius sections) through the blade. These foil shapes are known from propeller design. This approach was adopted by Umlauf [76], who has implemented a skinning representation of the propeller blade surface (as well as a fillet surface for the hub radius transition). He proceeds in the following major steps:

- Given propeller design data:

  $\phi(r)$ pitch angle distribution

  $c(r)$ chord length distribution

  $t(r)$ thickness distribution

  $m(r)$ camber distribution

  $R(r)$ rake distribution

  $S(r)$ skew distribution

  r radial coordinate

- Foil data:
  Data for two-dimensional foil sections for propellers are often derived from the NACA foil series based on thickness and camber distributions. Many families of wing sections (thickness distributions) and mean lines (camber lines) are published in [1]. For example, the NACA four digit and five digit wing sections have dimensionless thickness distributions defined by

$$\pm y_t = t_x/0.2 \left(0.29690\sqrt{s} - 0.12600s - 0.35160s^2 + 0.28430s^3 - 0.10150s^4\right) \tag{444}$$

where t_x is the maximum thickness of the section, and s the dimensionless chordwise coordinate: $0 \leq s/c \leq 1$.

The NACA $a = \cdots$ mean lines, which are often used in propeller design for their favourable cavitation performance, have a dimensionless representation of

$$
\begin{aligned}
y_m/c &= \frac{C_{Li}}{2\pi(a+1)} \left\{ \frac{1}{1-a} \left[\frac{1}{2}\left(a - \frac{s}{c}\right)^2 \ln\left|a - \frac{s}{c}\right| - \frac{1}{2}\left(1 - \frac{s}{c}\right)^2 \ln\left(1 - \frac{s}{c}\right) \right.\right. \\
&\left.\left. + \frac{1}{4}\left(1 - \frac{s}{c}\right)^2 - \frac{1}{4}\left(a - \frac{s}{c}\right)^2 \right] - \frac{s}{c}\ln\frac{s}{c} + g - h\frac{s}{c} \right\}
\end{aligned} \tag{445}
$$

where

$$g = \frac{-1}{1-a}\left[a^2\left(\frac{1}{2}\ln a - \frac{1}{4}\right) + \frac{1}{4}\right] \tag{446}$$

$$h = \frac{1}{1-a}\left[\frac{1}{2}(1-a)^2 \ln(1-a) - \frac{1}{4}(1-a)^2\right] + g \tag{447}$$

and $a = s/c$ is a knuckle point at which the load distribution changes from a uniform chordwise loading to a linearly decreasing loading. C_{Li} is the 'design' or 'ideal' lift coefficient. This is the lift coefficient which corresponds to the ideal angle of attack (such that infinite velocities are avoided at the leading edge [1]).

The thickness function coordinates are usually plotted perpendicularly to the mean line whose local slope angle is θ. The foil coordinates for the cambered section are then (Fig. 122):

$$
\begin{aligned}
X_U &= s - y_t \sin\theta \\
Y_U &= y_m + y_t \cos\theta
\end{aligned} \tag{448}
$$

for the suction side (upper), and:

$$
\begin{aligned}
X_L &= s - y_t \sin\theta \\
Y_L &= y_m + y_t \cos\theta
\end{aligned} \tag{449}
$$

for the pressure side (lower). To summarize the notation for foil thickness coordinates we introduce:

$$t(s) = \begin{cases} -Y_U & \text{for the suction side} \\ -Y_L & \text{for the pressure side.} \end{cases} \tag{450}$$

The minus sign accounts for the change from foil to propeller coordinates.

- Geometry of blade surface:

  The foil sections are placed on cylindrical sections of the blade and are attached to the generating line (spine curve) of the propeller. This results in the following relationships for points in the blade surface $P(x, y, z)$, where

  $x=$ horizontal distance to starboard from propeller origin

  $y=$ distance from propeller origin in negative advance direction

  $z=$ vertical distance above propeller origin

  $$
  \begin{aligned}
  x(r, s) &= -r \sin \omega(r, s) \\
  y(r, s) &= y_G(r) + (s - c(r)/2) \sin \phi(r) + t(s) \cos \phi(r) \\
  z(r, s) &= r \cos \omega(r, s)
  \end{aligned}
  \tag{451}
  $$

  with

  $$
  \begin{aligned}
  \omega(r, s) &= -x_D(r, s)/r \\
  x_D(r, s) &= x_G(r) + (s - c(r)/2) \cos \phi(r) + t(s) \sin \phi(r)
  \end{aligned}
  \tag{452}
  $$

  and with the coordinates of the generating line

  $$
  \begin{aligned}
  x_G(r) &= -S(r) \cos \phi(r) \\
  y_G(r) &= S(r) \sin \phi(r) + R(r) \\
  z_G(r) &= r.
  \end{aligned}
  \tag{453}
  $$

  These equations are sufficient to evaluate any desired point in the blade surface from given design data [69].

Skinning:

  It is therefore possible to generate sets of data points for a family of skin curves at constant radii. These data can be interpolated by B-spline curves, which will then serve as the starting point for the skinning procedure. The interpolation may be based on spline fairing to ensure smoothness of the curve set [76].

  The blade surface may be approximated by a single B-spline surface for the pressure and suction sides. It is also possible to represent the two sides by two separate surfaces with a continuity constraint at the nose. In any case a greater density of points is needed for accuracy reasons both near the leading and trailing edges. The sharp trailing edge, but also the lines where the nose radius blends into the sides should be made into knot lines. The B-splines functions should be at least cubic in both directions to achieve curvature continuity.

  The skinning interpolation strictly requires end conditions at the hub and tip radii. At the hub it is sufficient to let the curvature of the surface equal zero because a fillet is later faired in here anyway. The tip requires special consideration because the skin curve there reduces to a point. The skinning algorithm still works in this vicinity, but the resulting surface has a singular point at the tip where it cannot be evaluated. A special construction can be used to overcome this problem [76].

Figs. 121-125 show an example of the skinning representation of a propeller blade and projective views with isolines to illustrate the shape of the blade.

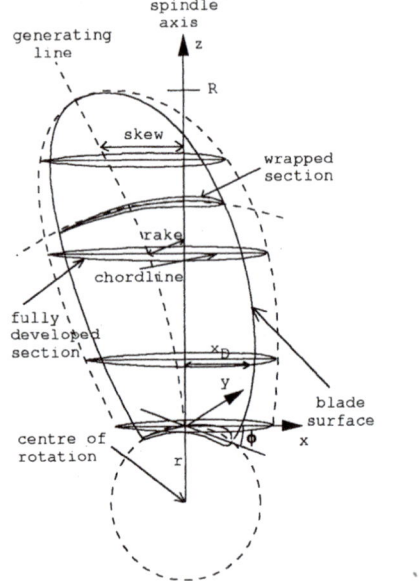

Fig. 121. Propeller geometry

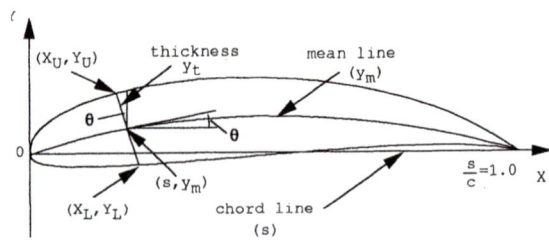

Fig. 122. NACA foil geometry

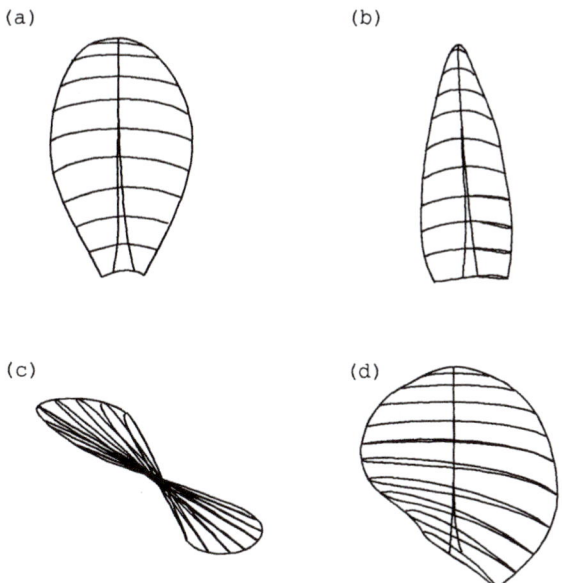

Fig. 123. Orthogonal projections a, b, c and perspective view d of propeller blade

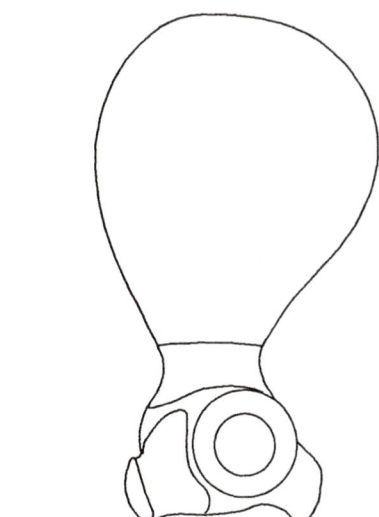

Fig. 124. Front view of blade with hub and fillet surface

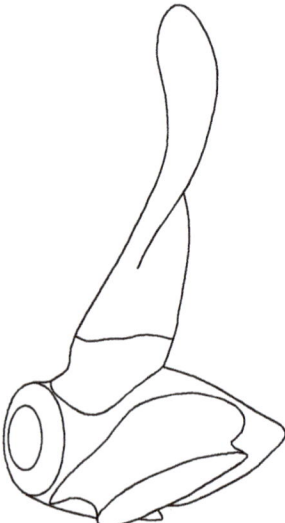

Fig. 125. Side view of blade [76]

5. Summary

Processes of surface generation were reviewed in terms of given data types, surface representations, criteria and constraints. This yields a classification for different types of surface generation problems whose common denominator can be regarded to lie in the format of optimization with constraints.

This section then reviewed surface representations of Coons Boolean sum, Cartesian product, Bézier and B-spline types. Elastic fairing processes and a few other, more specialized generation methods were also discussed. Finally the skinning approach to surface generation and its application to the representation of a propeller blade were presented to illustrate the generation of B-spline surfaces from given cross-sectional data.

CHAPTER 9
SHIP SURFACE DESIGN

1. Problem Statement

Ship surface design is a special application of the surface generation processes described in the previous section or of similar procedures. As discussed in Chapter 5, the geometric requirements for ship hull forms are generally related to curves so that curve design of ship lines usually precedes ship surface design. Only very few form parameters are directly related to the entire hull surface, e.g., underwater volume and centroid location.

Consequently the great majority of ship surface design methods progresses from curve design via curve mesh definition to surface generation. Therefore in this section we will place our main emphasis on methods of curve mesh fairing and surface interpolation. An example of direct surface design from integral form parameter constraints will also be given.

2. Curve Mesh Fairing

Large portions of the ship surface can be defined via orthogonal, regular meshes of curves, e.g., waterlines and sections or more generally longitudinal and transverse curves. The fairing of these curve meshes ought to precede the surface definition. In contrast to conventional naval architecture practice, which is based on fairing individual curves successively and iteratively (Chapter 5), there have been recent developments in computer-based systems to fair the entire set of mesh curves simultaneously.

Apparently it was Hosaka [31] who first proposed a method for fairing regular curve meshes. It is based on the analogy to an elastic system of orthogonal beams with spring supports at the knot points (Fig. 126). Only the corner points are fixed positionally, though free to rotate. This configuration will obviously only approximate certain discrete data and end constraints. It is best suited at the early stage of design when the given data are not yet accurate.

The elastic mesh consists of m by n curves, which form a regular mesh. The given data points $\mathbf{P}_{i,j}$ are attached to springs of stiffness α_{ij}. The stiffness of the beams representing the curves may be constant ($= EI$). This yields the following energy norm for the beams:

$$L_2 = \frac{1}{2}EI \left(\sum_{i=1}^{n} \int_{C_i} \left(D^2 \mathbf{r}_i(u) \right)^2 du + \sum_{j=1}^{m} \int_{D_j} \left(D^2 \mathbf{r}_j(v) \right)^2 dv \right) \qquad (454)$$

where

$$\mathbf{r}_i(u), \mathbf{r}_j(v) = \text{mesh curves in } u- \text{ and } v- \text{ direction,}$$
$$i = 1, \cdots, n \quad j = 1, \cdots, m.$$

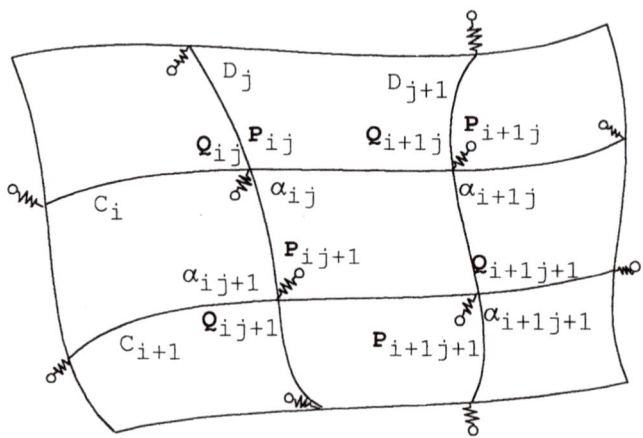

Fig. 126. Curve mesh with elastic splines and spring supports at the knots $\overline{P}_{i,j} = \mathbf{P}_{i,j}$; $\overline{Q}_{i,j} = \mathbf{r}(u_i, v_j)$

$$D^2 = d^2/du^2 \text{ or } d^2/dv^2$$

The distance constraint for all movable mesh knots $\mathbf{r}(u_i, v_j)$ can be written as an energy functional, too:

$$A = \frac{1}{2} \sum \sum \alpha_{ij} \left(\mathbf{r}(u_i, v_j) - \mathbf{P}_{i,j} \right)^2 - I_0 \leq 0 \qquad (455)$$

where I_0 is a positive error bound. This error bound as a measure of achievable and permissible tolerance can be determined from statistical considerations. Kaklis [35] has extended Hosaka's work in this direction and has suggested a range for I_0 based on the method of generalized cross-validation.

The free variational form corresponding to minimizing Eq. 454 subject to the constraint of Eq. 455 is defined by

$$I = L_2 + \lambda(A - d^2) = \min. \qquad (456)$$

This extreme value problems is solved numerically for the unknown curve coefficients, Lagrangian multiplier λ and slack variable d^2, normally from a nonlinear system of equations.

Hosaka's original method is a special case of this problem for $\lambda =$const., $d^2 = 0$ and I_0 very great. In this case the solution is much simplified and a linear system of equations results [53].

Figs. 127, 128 and 129 illustrate the effectiveness of this fairing process applied to a ship lines plan [54]. An initial distortion in the afterbody lines (Fig. 127) is removed by mesh fairing once (Fig. 128), then three more times (Fig. 129).

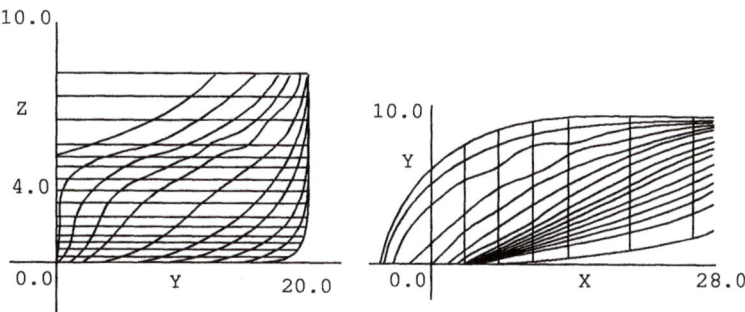

Fig. 127. Afterbody sections and waterlines, before fairing, distorted [54]

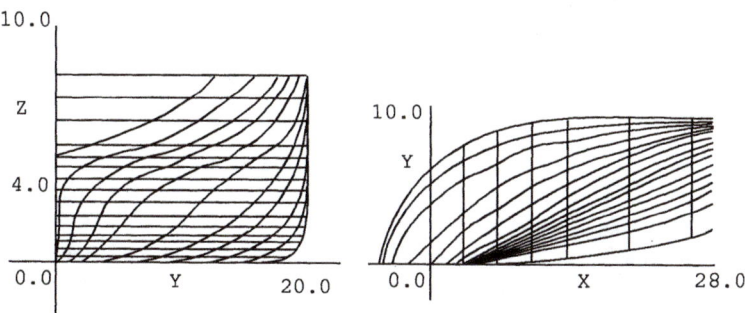

Fig. 128. Afterbody sections and waterlines, after fairing once [54]

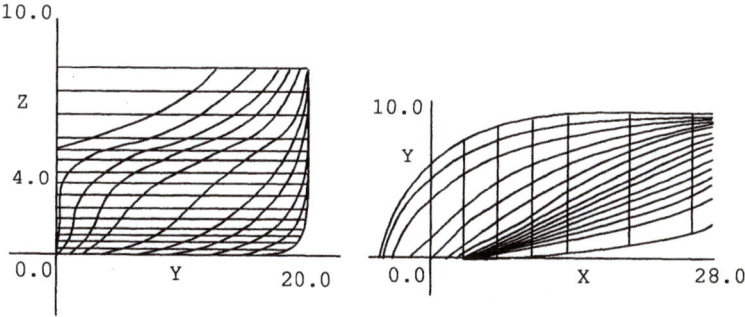

Fig. 129. Afterbody sections and waterlines, after fairing four times [54]

3. GC^1 Interpolation of Regular Meshes

The construction of a GC^1 continuous, i.e. tangent plane continuous, surface through a regular mesh of curves can be solved in two steps by first connecting two adjoining rectangular patches, then showing how to deal with the more general case of four patches surrounding one mesh knot. For the GC^1 connection of two adjoining Bézier patches several methods are known (Bézier [4], Hosaka and Kimura [32], Faux and Pratt [26], Farin [24]). Most of these methods furnish sufficient conditions for GC^1 surface construction. Liu and Hoschek [41] have derived necessary and sufficient conditions for solving this problem. We will follow their approach here (see also [54]).

Suppose the boundary curves in our regular mesh are cubic spline segments. Therefore it is possible locally to construct bicubic Bézier patches through four boundary curves and even to choose the interior vertices freely. However, a basic prerequisite for achieving GC^1 continuity between two adjoining patches is that one patch must be elevated in degree by one relative to the other. Let two neighbouring Bézier patches of degrees (s, m) and (s^*, l) have the representations

$$\mathbf{r}(u, v) = \sum_{j=0}^{m} \sum_{i=0}^{s} \mathbf{V}_{i,j} \cdot B_{i,s}(u) B_{j,m}(v) \tag{457}$$

$$\mathbf{r}^*(u, v) = \sum_{j=0}^{l} \sum_{i=0}^{s^*} \mathbf{V}_{i,j}^* \cdot B_{i,s^*}(u) B_{j,l}(v) \tag{458}$$

for $0 \leq u, v \leq 1$ with $s = n - 1, s^* = n$.

Let further the Bézier polyhedron edge vectors across the common boundary curve be denoted by $\mathbf{a}_i, \mathbf{a}_i^*$ and the polygon edge vectors along the common boundary by $\mathbf{e}_i$ (Fig. 130) as follows:

$$\left. \begin{array}{l} \mathbf{a}_i = \mathbf{V}_{i1} - \mathbf{V}_{i0} \\ \mathbf{e}_i = \mathbf{V}_{i+1,0} - \mathbf{V}_{i0} \end{array} \right\} \quad \text{for} \quad i = 0, \cdots, s,$$

$$\mathbf{a}_i^* = \mathbf{V}_{i1}^* - \mathbf{V}_{i0}^* \qquad i = 0, \cdots, s^*. \tag{459}$$

If the patch $\mathbf{r}(u, v)$ is constructed as a bicubic patch $(s = 3)$, then the $\mathbf{a}_i$ and $\mathbf{e}_i$ are known. The sufficient condition for GC^1 continuity between two patches yields the general rule

$$\mathbf{a}_i^* = \frac{i}{n} \lambda_1 \mathbf{a}_{i-1} + \left(1 - \frac{i}{n}\right) \lambda_0 \mathbf{a}_i + \frac{i}{n} \mu_1 \mathbf{e}_{i-1} + \left(1 - \frac{i}{n}\right) \mu_0 \mathbf{e}_i. \tag{460}$$

For tangent plane continuity the corner point tangent vectors of the prefaired mesh $\mathbf{a}_0, \mathbf{e}_0, \mathbf{a}_n^*$ and $\mathbf{a}_{n-1}, \mathbf{e}_{n-1}, \mathbf{a}_n^*$ must be coplanar. From Eq. 460

$$\begin{array}{rcl} \mathbf{a}_0^* & = & \lambda_0 \mathbf{a}_0 + \mu_0 \mathbf{e}_0 \\ \mathbf{a}_n^* & = & \lambda_1 \mathbf{a}_{n-1} + \mu_1 \mathbf{e}_{n-1} \end{array} \tag{461}$$

which serves to determine the four free coefficients $\lambda_0, \lambda_1, \mu_0, \mu_1$.

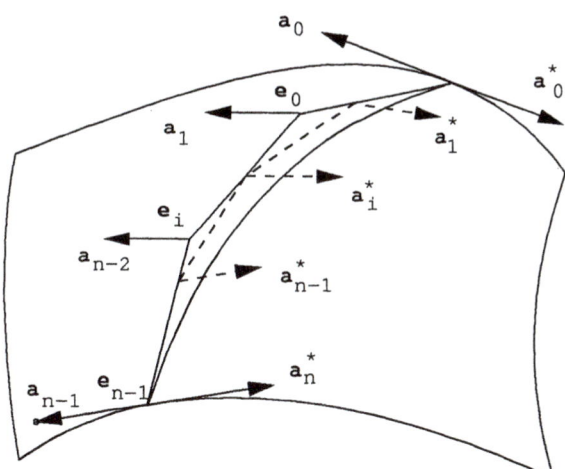

Fig. 130. Connection of two Bézier patches, GC^1 continuity, with degree elevation from $(n-1)$ to n

To join four patches contiguous to one corner in a GC^1 fashion this construction would not be sufficient because some of the interior Bézier points would have to comply with conflicting conditions from two boundaries. Therefore a different procedure is used here [54]. The surface is subdivided into a regular mesh such that bicubic Bézier patches border on degree elevated patches in a chessboard pattern (Fig. 131). The bicubic patches are filled in first. Then the holes are filled with 'biquartic' patches of a special kind, so-called Gregory-Bézier patches [28], [16] of the form:

$$\mathbf{r}_G(u,v) = \sum_{j=0}^{m} \sum_{i=0}^{n} \mathbf{V}_{i,j} \cdot B_{i,n}(u) B_{j,m}(v) \qquad (462)$$

with the special provision that the interior control points next to the corners are given by a rational expression from which they can be split up into two separate parts so that each part can be determined independently from the continuity requirements at its adjacent boundary. The form of these vertices is, e.g. [54]:

$$V_{1,1} = \frac{u V_{1,1}^1 + v V_{1,1}^2}{u + v}. \qquad (463)$$

This is illustrated in Fig. 132.

An application of this method to a surface of 2 by 4 patches in a ship bow is shown in Figs. 133 and 134.

Extensions of this methodology to GC^2 continuous surface interpolation of mesh curves are known. Most procedures require discouragingly high polynomial degrees. A relatively economical procedure was developed by Weber [79]. A systematic review and extension of this methodology was described by Ye [84].

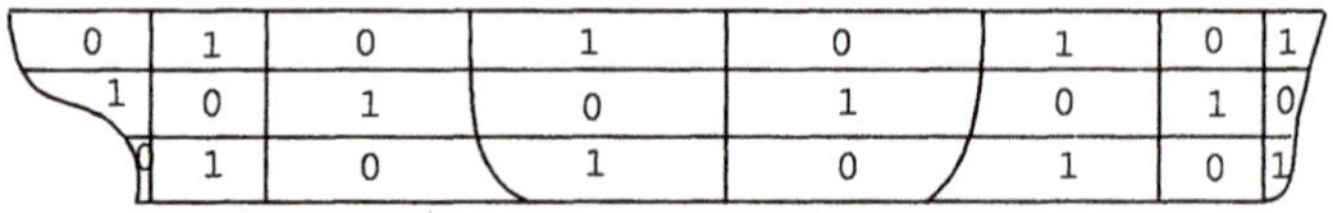

0	1	0	1	0	1	0	1
1	0	1	0	1	0	1	0
1	0	1	0	1	0	1

Fig. 131. Subdivision of ship surface into Bézier patches (label 1, degree $3 \cdot 3$ and Gregory-Bézier patches (label 0) [54]

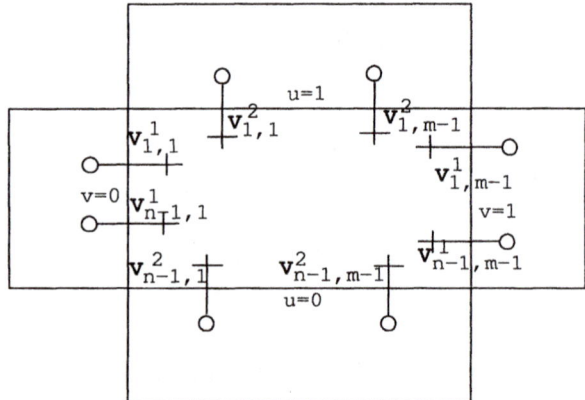

Fig. 132. Gregory-Bézier patch (degree $n \cdot m$) GC^1 connected to neighbouring Bézier patches (degree $(n-1) \cdot (m-1)$) [54]

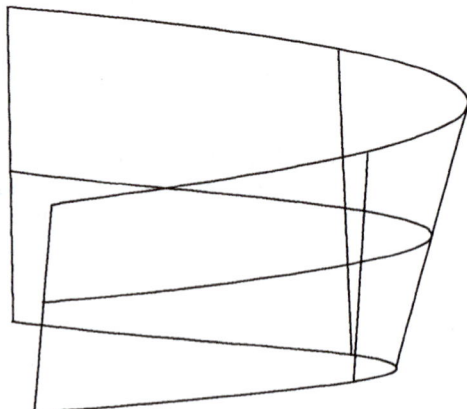

Fig. 133. Patch subdivision in ship bow, 2by 4 patches [54]

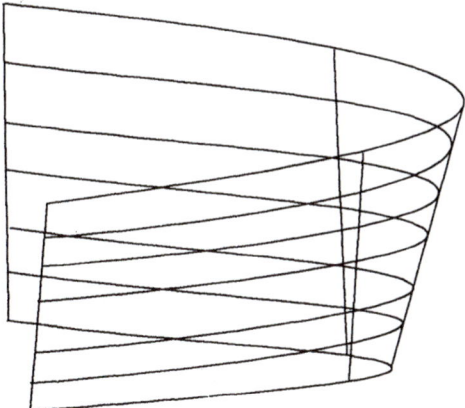

Fig. 134. Horizontal and transverse planar sections through ship bow panels [54]

4. Surface Design with Volume Constraints

The underwater characteristics are usually subject to constraints on volume and volume centroid location, hence integral property constraints of the surface, in addition to other constraints.

Standerski [71] has developed a method for constructing an underwater ship surface directly from such constraints. The problem is formulated as a variational problem in the following way:

Constraints are imposed on:

- Volume

- Volume centroid

- Waterplane shape (and area)

- Midship section shape (and area)

- Moment of inertia of the waterplane (I_T).

The hull is represented by a B-spline surface with a generous number of control vertices (say, 20 by 20).

The criterion function is Pilcher's criterion, a first order criterion [61]:

$$
\begin{aligned}
L_1 \;=\;& <X_u, X_u> + <X_v, X_v> \\
& + <Y_u, Y_u> + <Y_v, Y_v> \\
& + <Z_u, Z_u> + <Z_v, Z_v>
\end{aligned}
\tag{464}
$$

where the expressions $< X_u, X_u >$ etc. correspond to the integrated inner products of the corresponding first partial derivatives of the surface components X, Y, Z.

The minimization of this criterion tends to reduce the wetted surface of the hull form.

If the integral constraints are symbolically denoted to be of the type $F = 0$ then the Lagrangian free variational form of this optimization problem can be stated as:

$$I = L_1 + \nu F = \min. \tag{465}$$

This functional is minimized with respect to the unknown B-spline control polygon points, which provides the conditions for a system of equations to be solved numerically.

Standerski has presented a variety of examples [71]. Some results are shown in Figs. 135, 136.

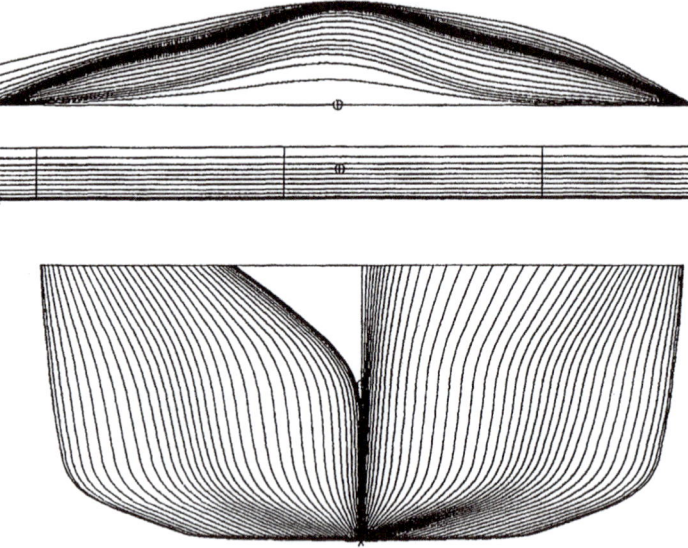

Fig. 135. Lines plan, created with volume constraints, Form parameters: $L = 141m, B = 22.3m, T = 6.0m, \nabla = 10000m^3, LCB = 69.2m, KB = 3.36m, KG = 9.0m, C_{WP} = 0.71, C_M = 0.92, GM/B = 0.024$

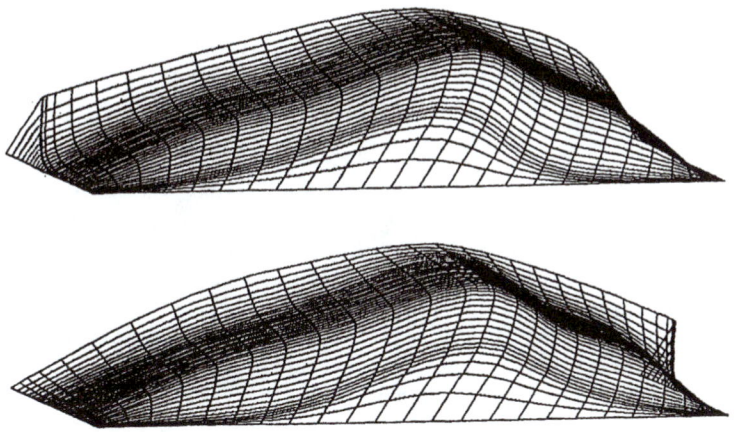

Fig. 136. Perspective view

5. Summary

The generation and design of ship surfaces is usually based on the prior development on ship lines which are conceived so that the principal functional properties and main shape characteristics of the hull form are secured. These lines act as 'soft' (or hard) constraints for the surface generation. The quality of the given lines is of crucial importance for the achievable quality of the surface.

Within the framework of this task a method for curve mesh fairing was presented in this section. The lines thus faired are then interpolated by parametric surfaces, usually at least in a tangent plane continuous way.

An example also illustrates the alternative of creating a ship surface directly from volume and centroid constraints with given boundary curves and end conditions, but without intervening curve mesh construction. This methodology is not yet mature, but merits further consideration because of its time-saving potential.

CHAPTER 10
ANALYSIS OF SURFACE FAIRNESS

1. Introduction

The development of a ship surface definition generally proceeds through two stages, design and fairing of the hull surface. During the design stage the hull is generated from ship curves which are in turn systematically faired and fitted as the ship is designed. However, it is observed that this mathematical fairing process remains laborious, mainly due to the complex fine tuning required on a mesh of numerous lines and data points.

This suggests that direct fairing of the ship surface, dealing with major surface patches simultaneously, may produce time-savings.

In previous chapters we have seen how the curve sets forming the hull form can be systematically faired, with techniques such as point removal, and manipulation of control vertices for B-spline curves. Then, to assess the curve fairness, the curvature distribution can be examined to see whether the curvature fluctuates from one side of the curve to the other, or whether there are large 'peaks' of curvature at any location. These can be removed by the various techniques described above to produce a smooth curvature distribution for each curve. Furthermore, by imposing constraints such as the minimization of the L_2 norm we can reduce the amount of fairing required for the curve sets. This is illustrated in Chapter 5, Section 3 with the design of a ship underwater section. It should be noted that for a section such as this the amount of fairing required will be minimal.

The fairing of surface patches needs different considerations. Firstly, the patches should tend to be large and mesh lines few for greater fairing effect. The patches can be created from a set of previously faired boundary curves, such as waterlines, sections or buttocks. However, fair boundary curves of the patches do not guarantee fairness of the patch interior. Therefore, the surface patch needs to be faired and analysed to obtain an overall picture of the fairness of the surface. In iterative fairing, effective visualization of shape characteristics and particularly fairness flaws on limited display screens is of crucial importance. Scale and resolution usually need to be considered when the fairness of a surface is judged, since variations in scale can appear to give different results.

However, the analysis of the fairness of a surface patch is a more difficult concept to grasp, than could be applied to the fairing of curves. Methods used in the car industry rely on the reflecting properties of light; the car is placed in a room with parallel fluorescent strip lights on the ceiling. These lights reflect in the polished car surface, producing so called 'reflection lines' whose own fairness governs the fairness of the surface. Thus, if there are 'wiggles' or imperfections in the lines then there is deemed to be a flaw in the car body.

This, however, does not provide us with a mathematical formulation for assessing the fairness of the surface. To do this we need examine the local properties of the surface, defined by the Gaussian and mean curvature of the surface.

2. Curvature Properties of a Surface

In Chapter 2 we saw that the curvature vector of a curve parametrized by arc length gave a measure of the twist of the tangent vector as it moved along the curve and was given by

$$t' = \kappa n = k. \tag{466}$$

If we consider this curve to be an isoline of the surface, then this vector k can be decomposed into a component k_n normal and a component k_g tangential to the surface, such that:

$$t' = k = k_n + k_g \tag{467}$$

where we term the vector k_n the normal curvature vector, and k_g the tangential or geodesic curvature vector [73], which will not be discussed in these notes.

Now the vector k_n can be expressed in terms of the unit surface normal n as

$$k_n = \kappa n \tag{468}$$

where the scalar κ is termed the normal curvature. This depends for its sign on the sense of the normal vector n, with the vector k_n being solely determined by the curve and not in any sense by the choice of t or n.

As $n \cdot t = 0$ we can differentiate this expression along the curve to obtain

$$\frac{dt}{ds} \cdot n = -t \cdot \frac{dn}{ds} = \frac{dx}{ds} \frac{dn}{ds} \tag{469}$$

which gives us an expression for the normal curvature:

$$\kappa = -\frac{dx \cdot dn}{dx \cdot dx}. \tag{470}$$

Since we know that both n and x are surface functions of u and v, we can, with the aid of the identities:

$$dn = n_u du + n_v dv, \quad dx = x_u du + x_v dv \tag{471}$$

rewrite Eq. 471 in the form

$$\kappa = -\frac{(x_u \cdot n_u)du^2 + (x_u \cdot n_v + x_v \cdot n_u)dudv + (x_v \cdot n_v)dv^2}{(x_u \cdot x_u)du^2 + 2(x_u \cdot x_v)dudv + (x_v \cdot x_v)dv^2}. \tag{472}$$

Furthermore, since $x_u \cdot n = 0$ and $x_v \cdot n = 0$, we can again differentiate along the curve to obtain the expressions

$$x_{uu} \cdot n = -x_u \cdot n_u, \qquad x_{vv} \cdot n = -x_v \cdot n_v$$
$$x_{uv} \cdot n = -x_u \cdot n_v \quad = \quad -x_v \cdot n_u = x_{vu} \cdot n \tag{473}$$

and so we obtain the expression for the normal curvature

$$\kappa = \frac{e\,du^2 + 2f\,du\,dv + g\,dv^2}{E\,du^2 + 2F\,du\,dv + G\,dv^2} = \frac{II}{I} \qquad (474)$$

where e, f, g, E, F, G are the fundamental coefficients of the first and second fundamental forms, derived in Chapter 6, and where κ is fully dependent on the directions
of dv/du.

In Fig. 137 the plane $\mathbf{P}$ cutting the surface $\mathbf{S}$ contains the surface normal $\mathbf{n}$ at the
point $\mathbf{A}$ on the surface. At the point $\mathbf{A}$, the line of intersection $\mathbf{C}$, has curvature κ.
As the plane $\mathbf{P}$ is rotated about the surface normal, the curvature κ varies. There are
certain unique directions for which the curvature κ reaches a maximum and minimum.
These can be derived mathematically.

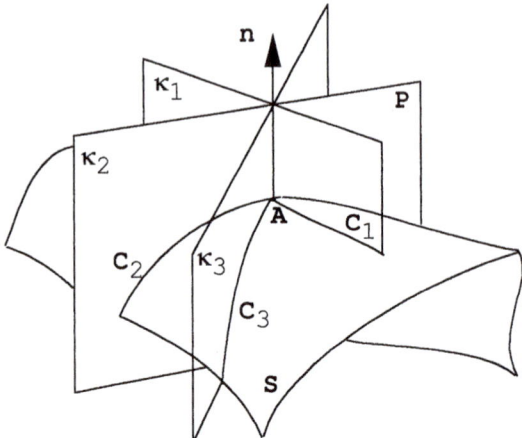

Fig. 137. The normal curvature at a point on the surface

The normal curvature in direction dv/du, given by Eq. 474 can be rewritten as a
function of $\lambda = dv/du$ such that

$$\kappa = \frac{e + 2f\lambda + g\lambda^2}{E + 2F\lambda + G\lambda^2} = \kappa(\lambda) \qquad (475)$$

from which the extreme values of $\kappa(\lambda)$ can be characterized by

$$\frac{d\kappa}{d\lambda} = 0 \qquad (476)$$

which implies that

$$(E + 2F\lambda + G\lambda^2)(f + g\lambda) - (e + 2f\lambda + g\lambda^2)(F + G\lambda) = 0. \qquad (477)$$

However, since

$$\begin{aligned}
E + 2F\lambda + G\lambda^2 &= (E + F\lambda) + \lambda(F + G\lambda) \\
e + 2f\lambda + g\lambda^2 &= (e + f\lambda) + \lambda(f + g\lambda)
\end{aligned} \qquad (478)$$

we can write Eq. 475 in the form

$$\kappa = \frac{II}{I} = \frac{f + g\lambda}{F + G\lambda} = \frac{e + f\lambda}{E + F\lambda} \qquad (479)$$

and hence κ satisfies

$$(e - \kappa E)du + (f - \kappa F)dv = 0, \quad (f - \kappa F)du + (g - \kappa G)dv = 0 \qquad (480)$$

which on elimination of κ gives a quadratic equation for λ with two real roots:

$$(Fg - Gf)\lambda^2 + (Eg - Ge)\lambda + (Ef - Fe) = 0 \qquad (481)$$

which determines the directions dv/du in which κ obtains its extreme values, the maximum and minimum values. These directions are known as the directions of principal curvature or curvature directions [73].

Furthermore, since the roots λ_1, λ_2 of the quadratic (Eq. 481) satisfy the equation

$$G\lambda_1\lambda_2 + F(\lambda_1 + \lambda_2) + E$$

$$= \frac{-1}{gF - Gf}[G(eF - Ef) - F(eG - Eg) - E(gF - Gf)] = 0 \qquad (482)$$

we can note that the curvature directions are orthogonal, as described in [73].

The normal curvatures in the curvature directions are termed the principal curvatures, and are denoted by κ_1 and κ_2. The lines of curvature on the surface are obtained by integration of Eq. 481 which form two sets of curves intersecting at right angles, or an orthogonal family of curves on the surface.

Now, since λ_1 and λ_2 satisfy Eq. 481 we can find the values of the principal curvatures κ_1 and κ_2 by substituting into Eq. 480. A faster method is found by observing that the values of κ_1 and κ_2 satisfy the two equations

$$\begin{aligned}
(E\kappa - e) + (F\kappa - f)\lambda &= 0, \\
(F\kappa - f) + (G\kappa - g)\lambda &= 0
\end{aligned} \qquad (483)$$

which can be simultaneously satisfied if and only if

$$\begin{vmatrix} E\kappa - e & F\kappa - f \\ F\kappa - f & G\kappa - g \end{vmatrix} = 0 \qquad (484)$$

which has two roots, κ_1 and κ_2. From this equation we can derive the two important quantities

$$H = \frac{1}{2}(\kappa_1 + \kappa_2) = \frac{Eg - 2fF + gG}{2(EG - F^2)} \tag{485}$$

which is termed the mean curvature, and

$$K = \kappa_1\kappa_2 = \frac{eg - f^2}{EG - F^2} \tag{486}$$

which is the Gaussian curvature.

3. The Distribution of Gaussian Curvature

The Gaussian curvature can now easily be calculated for a parametrized surface patch. This can be used to compare the shape of the surface. If the Gaussian curvature at a point, κ, > 0 then the surface is elliptic (compared with the same properties of the second fundamental form). If $\kappa < 0$ then the surface is hyperbolic, and if $\kappa = 0$ then the surface is either parabolic or locally flat.

To show the shape of a region we can plot the distribution of κ over the surface. Standard procedures are to plot the contours of Gaussian curvature. Thus, if a graph of Gaussian curvature shows that $\kappa > 0$ everywhere over a continuously curving region, then the entire region is elliptic, and similarly for the cases, $\kappa < 0$ and $\kappa = 0$.

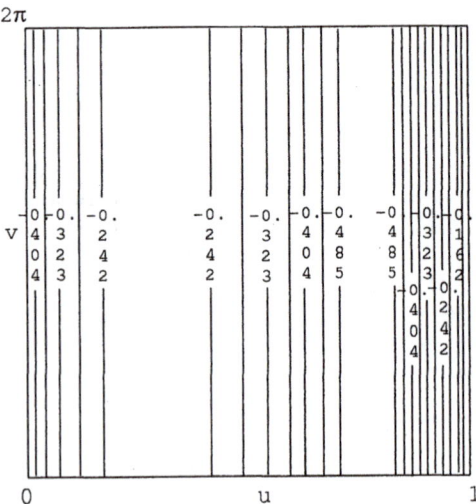

Fig. 138. Gaussian curvature over a surface patch

Fig. 138 illustrates a standard plot of Gaussian curvature over a (u,v) surface patch. It can be seen that the surface can be regarded as fair since the lines of constant curvature are straight and no surface distortions are present [48]. From the values associated with the lines of constant Gaussian curvature we can further see that the surface patch is hyperbolic everywhere. The figures reproduced throughout this chapter are reproduced from work of Brown [11].

It is less obvious how a pattern of Gaussian curvature isolines must be interpreted in terms of fairness. If we take the Gaussian contours of Fig. 138 which illustrate the smooth characteristic of the surface and apply a slight depression at a point, we see that the Gaussian curvature contour plot will now be as illustrated in Fig. 139.

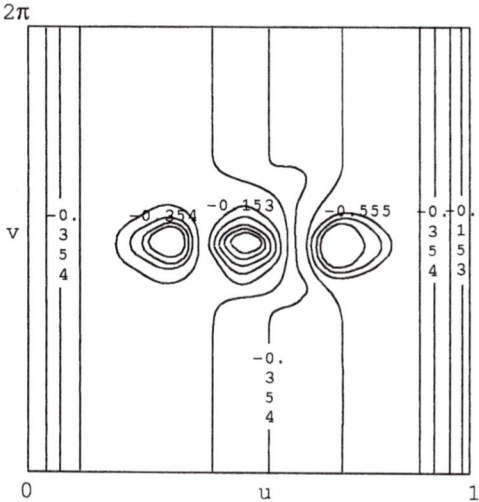

Fig. 139. Gaussian curvature for a surface with a slight depression

It can be seen that at the edges of the patch far from the depression the surface remains fair, however as the area of the depression is approached, the contours become closer and form closed curves on the surface patch. It should be noted that depending on the scaling used, these surface bumps could be termed fair, and it is therefore important to ensure an appropriate length scale is used for visualization.

The surface patch used in the previous example corresponds to a surface blend between a cylinder and a flat plane; the generation of which was described in Chapter 7. By inclining the plane and blending between the cylinder and the plane, as may be required between an internal beam and a ship's hull, the Gaussian curvature over the surface patch will alter dramatically. The Gaussian curvature is plotted in Fig. 140.

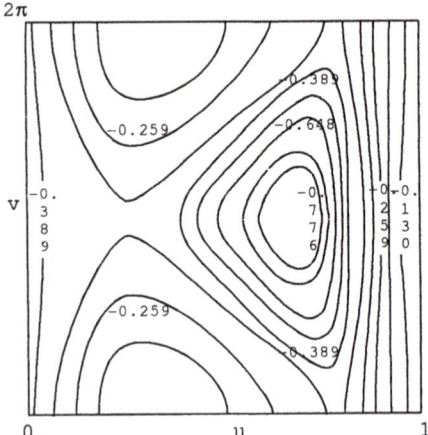

Fig. 140. Gaussian curvature on a surface blend

It is observed that a symmetric contour pattern is produced with a concentration of Gaussian curvature appearing close to the parameter value $u = 1$ where the blend is more highly curved. This is represented by the closed curves which attain a minimum curvature of -0.778.

Again creating an indentation on the surface affects the plot of Gaussian curvature, as can clearly be seen from Fig. 141.

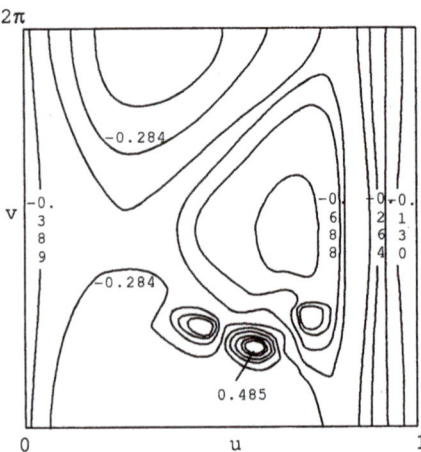

Fig. 141. Gaussian curvature on the modified surface

Further to the Gaussian contours being of a closed form in the region of the depression, the lines appear disjoint and 'wiggles' appear in the contours. These wiggles, as observed by Munchmeyer [48], illustrate the local fairness of the surface. If the wiggles can be removed by changing the length scale on the surface, then the surface may be fair; however if these wiggles persist, then the surface is considered to be unfair.

A similar interpretation of curvature distributions can be applied to the surface of a ship body plan. If the Gaussian curvature on a ship body appears well ordered we may take the surface to be fair; however should local depressions or wiggles occur then a fairing of the surface may be necessary. This we may undertake by methods described in Chapter 9.

4. Mean Curvature

The mean curvature H is the arithmetic mean of κ_1 and κ_2. Whereas Gaussian curvature tells whether a region is elliptic, hyperbolic or parabolic, the mean curvature indicates whether the region is full or hollow. In practise the Gaussian and mean curvature should both be used to determine the local shape of the surface.

This can be illustrated by considering the case where the Gaussian curvature is greater than zero, and so the surface is elliptic. To determine whether the surface is full or hollow we must regard the mean curvature. A local protrusion or bump will have $H > 0$ while an indentation will have $H < 0$, with the magnitude of H indicating the sharpness of the bump or hollow, as can be seen in Fig. 142.

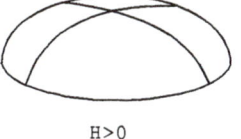
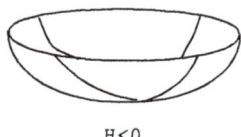

H>0   H<0

Fig. 142. The mean curvature on a surface

5. Colour Coded Representation

Although we have a way of illustrating the Gaussian and mean curvature of a surface patch, problems may arise when trying to interpret the diagrams. Reasons for this may occur due to the numerical values on the contours being too crowded and hard to distinguish on a patch with many contours. Alternatively, the contour lines may be subject to misinterpretation by a designer initially using the techniques.

Thus, a useful method for displaying the curvatures, is by colour coding the values of the curvatures and using these values as a quantitative measure of a fourth dimension on the actual three dimensional surface [22]. Using this method potential problem areas of surface fairness are made immediately apparent to the designer, in a single image. Therefore, this approach may improve design quality while reducing the time required. One such example of the mean curvature over the surface is illustrated below in Fig. 143 in grey-scale shading, where a blend between two cylinders at right angles to each other is considered.

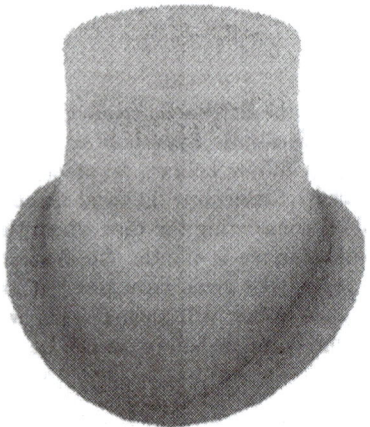

Fig. 143. Grey-scale representation of mean curvature

CHAPTER 11

HYDRODYNAMIC EVALUATION
OF GENERATED SURFACE

1. Introduction

During the process of ship design a naval architect tries to create a ship body form that will meet the desired engineering, economical and functional requirements. For instance, these can be the task of a body creation with a fixed value of displacement, the problem of designing a ship with a chosen level of stability or to find a body form with good resistance features. In particular the last problem is, on the one hand, very important due to its economical aspect and, on the other hand, difficult to evaluate based on theoretical methodology. Since the time of W. Froude's experiments and his proposition of a model-ship resistance extrapolation algorithm, based on the phenomenological methodology, in shipbuilding practice this problem has mainly been solved by experiments in model tanks.

In the scope of these types of problems we have several interesting, engineering tasks. Among them are:

- Wave resistance and viscous resistance evaluation of a known ship body form and given velocity (total resistance prognosis),

- searching for a ship form with minimum resistance (and fulfilling some side conditions),

- local corrections of a given ship form to improve resistance characteristics (bulbous bow design procedure),

- analysis of flow stream lines geometry on the the body surface,

- evaluation of fluid velocity distribution of the wake profile in the stern region, etc..

For a very long time, engineering analysis of the above problems, as a part of a design process, was impossible to perform based on the rational-mechanics theoretical methods because of the complicated character of such calculations. The computer era opened possibilities of such analysis using proper theories and algorithms; the methods of which are still under development and improvement. Some authors term this set of methods 'computational towing tank'. All of these tasks can be treated in many analytic ways: as a straight geometrical approach or by hydrodynamical singularities modelling approach; using different theories - nonlinear or linearized; using distinct algorithms and numerical procedures. Some of these combinations give different design methods.

In this chapter we will discuss the theoretical relation between the geometry of a ship body surface and the wave resistance generated by this form. In particular, we will focus our attention on the evaluation of a ship surface, modelled by a set of surface patches of a chosen class, from a point of view of generated theoretical wave resistance which is based on thin ship wave resistance theory [46], [37]. Furthermore the surface patches are described by parametric equations. Due to their universality, surface patch techniques for the geometrical description of engineering objects are almost obligatory in modern technology. Thus, it is interesting to show their application in modern ship design methodology instead of a classical description of a ship body surface by explicit equations.

Attempts to treat the ship resistance problem theoretically have a long history. Since Michell developed his well known thin-ship wave resistance theory, the theory has been applied to numerous ship design problems, mostly with mixed results. This is mainly due to the limitations inherent in the theory. The theory is only valid under the assumptions:

- that the free-surface disturbances created by a moving ship are small in comparison to wave length,

- that the ratio B/L is small,

- that the viscosity effect is small.

For most practical ship forms, the free-surface disturbances are not necessarily small, not all ships are thin, and the viscosity effect, especially in the afterbody, is not negligible. Hence, it is optimistic to hope that thin-ship theory could be applied to the design problem in general. We have chosen this theory as the first step in the evaluation of a ship geometry, because it is relatively simple and gives us a straightforward relationship between the body shape and the evaluation of the generated wave resistance.

Efforts to improve the theory are being made by many researchers in this field. However, it will probably be a long time before a generally valid wave resistance theory can be developed. In the meantime, it is our belief that benefit can be obtained by using existing theory within its limitation.

2. Mathematical Model of Ship Wave Resistance

The physical phenomenon under consideration is that of a flow around the ship body moving forward at a constant speed on a surface. Due to gravitational surface waves, produced by the hull movement, the ship experiences wave resistance. This resistance is related to the hull shape.

The foundations of the ship wave resistance theory include some simplifying assumptions concerning the physical characteristics of the fluid, the boundary conditions and the kinematic features of the flow. Wave resistance of a ship, being a part of total resistance, is caused by a nonuniform pressure field around the moving ship hull. The value of wave resistance the ship experiences depends on:

  the density of water,

- the actual speed of the ship,

- the shape of the ship hull,

- the dimensions of the water reservoir.

Here, the fluid is assumed to be a Newtonian ideal fluid and the thermal phe-
nomena in it are ignored. Additionally, the flow is postulated to be irrotational and
stationary; the ship is thin and the dimensions of sea are infinite.

In the theoretical model of the potential flow the total resistance of a ship is equal
to wave resistance. Due to the existence of a free surface d'Alembert's paradoxon
does not apply.

With all the assumptions made above the ship wave resistance determined by thin
ship wave resistance theory is expressed by a formula referred to in the literature as
the Michell integral [46].

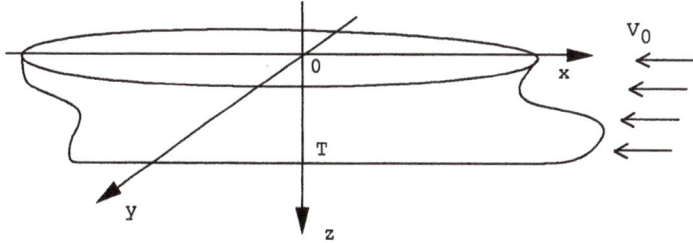

Fig. 144. The coordinate system connected with the ship hull. Plane XOY is equivalent to the
undisturbed water surface. A homogeneous liquid stream flows over the still hull from direction
$x = \infty$ with a speed of v_0

For the coordinate system as in Fig. 144 and for a surface of a ship body form
described by an explicit equation $u = u(x, z)$ the wave resistance can be expressed by
the Michell formula [37]

$$R_w = \Phi(u) = \frac{4\rho g^2}{\pi v_0^2} \int_0^{\pi/2} (H_s^2 + H_c^2) \frac{d\Theta}{\cos^3 \Theta} \qquad (487)$$

where

$$\left.\begin{array}{c} H_s \\ H_c \end{array}\right\} = \iint_\Omega \frac{\partial u(x, z)}{\partial x} \exp\left(\frac{-gz}{v_0^2 \cos^2 \Theta}\right) \left\{\begin{array}{c} \sin \\ \cos \end{array} \left(\frac{gx}{v_0^2 \cos \Theta}\right) d\Omega\right.$$

$$= \iint_\Omega \frac{\partial u(x, z)}{\partial x} \left\{\begin{array}{c} \phi(x, z; \Theta) \\ \psi(x, z; \Theta) \end{array} d\Omega\right. \qquad (488)$$

The given flat domain $\Omega \subset R^2$ is a part of the ship symmetry plane, XOZ, in
which the function u is determined.

The functional $\Phi : u \in D(\Phi) \rightarrow R \geq 0$ maps the admissible function space into R^+, a space consisting of the non-negative part of the Euclidean number axis. The function $u : \Omega \subset R^2 \rightarrow R$ is a function whose graphic representation is expressed by the vector function $\mathbf{r} : \Omega \subset R^2 \rightarrow R^3$ and which describes the ship hull surface, i.e.

$$\mathbf{r} \equiv \text{ graph } u \equiv \{(x, u(x, z), z) \subset R^3 : (x, z) \in \Omega\}. \tag{489}$$

3. Discretization by Surface Patches

The simplest mathematical model of a surface $S \in R^3$ is a patch. A patch $\mathbf{r}$ is thought of as a set of points bounded by a certain closed curve, the Cartesian coordinates of the points being expressed by continuous single-valued functions of two independent variables (parameters):

$$x = x(\xi, \eta), \quad y = y(\xi, \eta), \quad z = z(\xi, \eta); \quad x, y, z : R^2 \rightarrow R. \tag{490}$$

Usually, for the sake of convenience, the variables ξ and η belong to the interval $[0,1]$ or $[-1,1]$. Assuming that one of the independent variables is constant, one obtains a curve lying on the patch. For successive constant values substituted for this variable one obtains a family of curves lying on the patch. Similar procedures applied to the second variable yield another family of curves, which cross the first family of curves in such a way that only one curve from each family passes through an arbitrary point on the patch. The curve making up the boundary of the patch and satisfying the above definition, consists of four lines joined together in such a way that they form four corners referred to as the nodes of the patch.

The problem of conditions which should be satisfied by a function describing a patch is closely related to the construction of the patch. Assuming that the domain of determination of the parameters ξ and η is the Cartesian product of the intervals $\Omega \equiv [0,1] \times [0,1]$, then the coordinates of the patch nodes can be expressed as $\mathbf{r}(0,0), \mathbf{r}(1,0), \mathbf{r}(0,1), \mathbf{r}(1,1)$ and the boundary lines can be expressed as functions of one of the two independent variables: $\mathbf{r}(\xi, 0), \mathbf{r}(1, \eta), \mathbf{r}(\xi, 1)$ and $\mathbf{r}(0, \eta)$.

Under the above assumptions a patch can be defined as a set

$$S \equiv \{\mathbf{r}(\xi, \eta) : R^2 \rightarrow R^3; \quad (\xi, \eta) \in \Omega\}. \tag{491}$$

Using the concept of a patch one can consider a surface as a set of patches. These patches should satisfy a specified set of boundary conditions and also satisfy certain geometrical conditions while passing from one patch to the adjacent one, such as tangent plane continuity. These may be referred to as interface conditions.

Let $\mathbf{r}(\xi, 0), \mathbf{r}(1, \eta), \mathbf{r}(\xi, 1), \mathbf{r}(0, \eta)$ denote prescribed boundary curves of a patch, as shown in Fig. 145.

In order to determine the surface of the patch one has to determine a function $\mathbf{r}$ which satisfies the assumptions and which is equivalent to the respective boundary curves for the parameters $\xi = 0, \xi = 1, \eta = 0$ and $\eta = 1$.

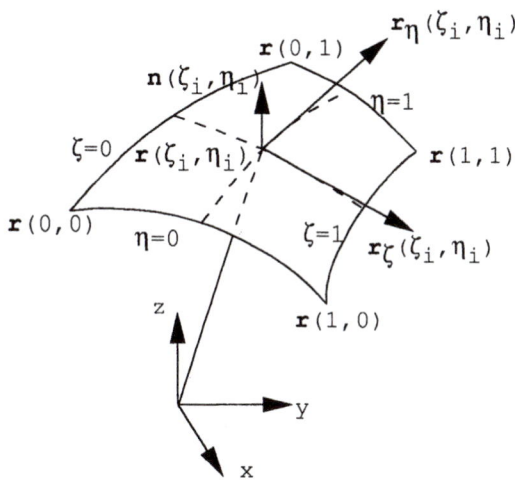

Fig. 145. Surface patch, notations

It is possible to show [26] that a function which fulfills the above assumptions can be expressed by the following vector function

$$\mathbf{r}(\xi,\eta) = \begin{bmatrix} \alpha_0(\xi) & \alpha_1(\xi) & \beta_0(\xi) & \beta_1(\xi) \end{bmatrix} \begin{bmatrix} \mathbf{r}(0,0) & \mathbf{r}(0,1) & \mathbf{r}_\eta(0,0) & \mathbf{r}_\eta(0,1) \\ \mathbf{r}(1,0) & \mathbf{r}(1,1) & \mathbf{r}_\eta(1,0) & \mathbf{r}_\eta(1,1) \\ \mathbf{r}_\xi(0,0) & \mathbf{r}_\xi(0,1) & \mathbf{r}_{\xi\eta}(0,0) & \mathbf{r}_{\xi\eta}(0,1) \\ \mathbf{r}_\xi(1,0) & \mathbf{r}_\xi(1,1) & \mathbf{r}_{\xi\eta}(1,0) & \mathbf{r}_{\xi\eta}(1,1) \end{bmatrix}$$

$$\begin{bmatrix} \alpha_0(\eta) \\ \alpha_1(\eta) \\ \beta_0(\eta) \\ \beta_1(\eta) \end{bmatrix} = \mathbf{F}(\xi)\mathbf{Q}\mathbf{F}^T(\eta). \tag{492}$$

The above form of a surface patch is usually referred to as the tensor product patch or the Cartesian product patch (Chapter 8, Section 3.2) and is a discrete structure, because the knowledge of the discrete values of the parameters of the patch at its nodes is sufficient for the unique determination of the structure.

The surface patch equation derived here can be made more specific if the interpolating functions $\alpha_0, \alpha_1, \beta_0, \beta_1$ are assumed to have the form of Hermite's polynomials. Such a patch preserves the property, that the interface condition is C^1 continuous on the boundary between adjacent patches, provided that the parameters in the matrix $\mathbf{Q}$ are properly selected. Every patch is defined by the values determined at four nodes, in which the four vectors are prescribed: $\mathbf{r}, \mathbf{r}_\xi, \mathbf{r}_\eta, \mathbf{r}_{\xi\eta}$, each vector having three components.

The choice of the description of a surface by parametric equations is not accidental, because this method is more advantageous than the description by functions expressed explicitly or implicitly. The advantages are as follows:

- Parametric equations provide more degrees of freedom for modelling the shape of a surface. For instance, the two-dimensional curve expressed in the explicit form

$$y = ax^3 + bx^2 + cx + d \qquad (493)$$

  has four parameters which can be used for modelling the curve. Applying the same class of functions and parametric equations the following relations are obtained

$$\begin{aligned} x &= au^3 + bu^2 + cu + d \\ y &= eu^3 + fu^2 + gu + h. \end{aligned} \qquad (494)$$

  There are eight parameters in them, which considerably extends the class of curves which can be described analytically; and $u \in R$ is the independent variable. For example, the problem of analytical description of the surface of the ship hull with a tunnel stern requires the application of relations of this type.

- It is not difficult to describe analytically hulls with a flat bottom or a transom stern (the surface of a hull perpendicular to its plane of symmetry), because an infinite derivative, e.g. $\frac{dy}{dz} \to \infty$ is obtained as follows

$$\frac{dy}{dz} = \frac{\frac{dy}{du}}{\frac{dz}{du}}, \qquad \text{where} \qquad \frac{dz}{du} = 0. \qquad (495)$$

- Owing to the normalization of independent variables, an automatic equivalence between the domain considered and the domain of a function is obtained.

- The geometrical properties of a surface are easy to express by vector quantities enabling the matrix notation of existing relations, which considerably facilitates coding a computational algorithm.

It is possible to (further) specify the approximation by the assumption of the following concepts concerning both the discretization of the function and its domain [57]:

- The domain of the function $\bar{\Omega}_u \subset R^n$ with the boundary $\partial \Omega_u$ is divided into E subdomains $\bar{\Omega}_u^e \subset R^n$.

- Each element $\bar{\Omega}_u^e$ is closed $\bar{\Omega}_u^e \equiv \Omega_u^e \cup \partial \Omega_u^e$ and includes a nonempty interior Ω_u^e.

- $\bar{\Omega} = \bigcup_{e=1}^{E} \bar{\Omega}_u^e.$

- $\Omega_u^e \cap \Omega_u^f = 0, e \neq f,$ where $e, f = 1, 2, \ldots, E.$

Then, in the case $n > 1$, the shape of elements is selected. In the case considered here with $\Omega \subset R^2$, elements in the form of convex triangles or quadrilaterals with linear or curvilinear sides are selected.

The application of curvilinear elements is justified in the case of complicated boundaries of the domain Ω because this allows the avoidance of a considerable reduction of the domain when a relatively small number of elements is employed. This, however, usually implies the necessity of transforming the domains to standardize the procedures of numerical integration.

There are two reasons in the problem considered here which made the use of curvilinear elements profitable. The first reason is the complicated shape of the domain Ω in the region of the bow (bulbous bow) and the stern. The second reason concerns the fact, that in the case of a surface describing the ship hull it is necessary to divide the surface into specific patches, e.g. a cylindrical part whose domain of determinacy is not the domain with linear boundaries. The selection of quadrilateral domains is due to the fact that quadrilaterals allow the construction of a C^1 continuous surface.

The generalized approximation coefficients a_i are assumed to represent the analytical properties of the function at the selected points of domains, the points being called nodes. In the problem considered here, these parameters take on the value of the function at the node $u(x_l, z_l)$ - the value of its first derivatives $u_x(x_l, z_l), u_z(x_l, z_l)$ and the value of the mixed derivative $u_{xz}(x_l, z_l)$, which suffices to construct a C^1 continuous surface. The corners of the domains are taken as the nodal points. On each element the function describing the ship surface is approximated by the following linear combination of basis functions

$$u_N^e = \sum_{i=1}^{N} \alpha_i^e \, \phi_i^e \, (x, z), \quad (x, z) \in \Omega_u^e \qquad (496)$$

where the coefficients α_i^e are the parameters of the approximating function at the nodes, and $\phi_i^e(x, z)$ are the basis functions of approximation.

The functions ϕ_i^e form a local basis of approximation, which is valid only on the domain $\bar\Omega_u^e$. These functions, apart from being linearly independent, are orthogonal to one another on the element and, moreover, they have a compact support, that is, they are identically equal to zero outside their domain of determinacy. The property of the basis functions allows the construction of the global solutions on the basis of partial local solutions, with the basis functions of the global solution preserving linear independence.

It is natural to choose bicubic Hermite polynomials as the local basis functions, with the coefficients satisfying the imposed requirements. Since each element $\bar\Omega_u^e$ can have a different shape due to nonlinearity of its boundary (preserving identical topology), integration of functions over various domains would be inconvenient. In order to avoid this, the domain of each element is transformed into a standard element of prearranged simple geometry. To avoid singular transformations, a standard element of the same topology is used - in the case in question it is a square. If the form of functions transforming the geometry of a domain has the same structure as the approximating function defined on this domain, then such transformations are called isoparametric.

However, application of nonlinear transformations causes the integrand to be multiplied by the Jacobian of the transformation and, moreover, functions which are linear before the transformation become nonlinear. In the problem considered here, this implied the development of special quadratures, which make it possible to obtain sufficient accuracy of integration, because the integrand of the Michell integral is a multioscillating function and thus difficult to integrate [45].

4. Isoparametric Transformations

According to Strang [72], there are two essential reasons which justify the application of the isoparametric transformation:

- If the domain Ω_u is a domain of irregular boundaries, the partition of Ω_u into the elements Ω_u^e induces a considerable approximation error in the case when the boundaries of elements $\partial\Omega_u^e$ are linear boundaries (domain reduction error). In order to decrease this error, the so called curvilinear elements Ω_u^e having boundaries $\partial\Omega_u^e$ that piecewise interpolate the boundary $\partial\Omega_u$ are used. Polynomials of suitable degree are usually employed as the interpolating functions.

  There is an essential requirement, that the elements being transformed from the global system to the local system (and vice versa) should preserve the required order of continuity on the boundaries of the elements. This condition determines the kind of transformation and, in particular, the degree of polynomials of the transforming function, and the kind of parameters of the transformation. In the case of the isoparametric transformation, the mapping $\Omega_u^e \leftrightarrow \Omega^*$ is based on the same set of nodal points with the same type of interpolation parameters as in the case of the interpolating function u^*, and thus the mapping is performed using the identical basis functions. This ensures that before and after the transformation the continuity condition is of the same order [85].

- The application of the mapping $\Omega_u^e \leftrightarrow \Omega^*$ unifies the domain of integration over individual elements Ω_u^e, thus reducing integrals over an element to iterated integrals of constant limits of integration being determined on the standard element Ω^*. However, the integrand is changed and after the transformation it is expressed as the composite function:

$$\tau : f[\, x, \quad z, \quad u(x,z) \,] \rightarrow F\{\, x(\xi,\zeta), \quad z(\xi,\zeta), \quad u[x(\xi,\zeta),z(\xi,\zeta)] \,\} \qquad (497)$$

  multiplied by the respective Jacobian of the transformation which represents the deformation of the areas of the domains being transformed.

If both functions f and F are easy to integrate numerically, then such an approach has essentially only advantages, except obviously, for the time required to develop suitable software. Nevertheless, the function F may happen to be more difficult to integrate than the original integrand f, which makes it necessary to apply special methods of numerical integration, or to increase the number of iterations in classical methods in order to obtain results with an acceptable error, provided that the time of computations is within the admissible limit.

This is the cost of convenient integration within constant iterated limits, which, on the other hand, facilitates the development of computer software.

The domain Ω_u^e given in the global system of coordinates XOZ is mapped into the domain Ω^* in the local system of coordinates $\zeta O \eta$, as in Fig. 146, using the following transformations

$$\tau^e : \Omega_u^e \to \Omega^*; \quad \tau^e = \left(\tau_x^e, \tau_z^e\right)$$

$$\tau_x^e : x^e \to x^e(\xi, \zeta; \ x_{k,l}^{m,n}) = \sum_{k=0}^{1}\sum_{l=0}^{1}\sum_{m=0}^{1}\sum_{n=0}^{1} x_{k,l}^{m,n} \ \Psi_{k,l}^{m,n}(\xi, \zeta)$$

$$\tau_z^e : z^e \to z^e(\xi, \zeta; \ z_{k,l}^{m,n}) = \sum_{k=0}^{1}\sum_{l=0}^{1}\sum_{m=0}^{1}\sum_{n=0}^{1} z_{k,l}^{m,n} \ \Psi_{k,l}^{m,n}(\xi, \zeta) \tag{498}$$

where $x_{k,l}^{m,n}$ and $z_{k,l}^{m,n}$ denote the nodal parameters of the domain Ω_u^e, and $\Psi_{k,l}^{m,n}(\xi, \zeta)$ denote the basis functions.

With the above mappings a question arises: how will a functional of the type $\iint_\Omega \frac{\partial}{\partial x} F(\xi, \zeta) d\Omega$ be transformed? It follows from the rules of differentiation of a composite function that

$$\frac{\partial F}{\partial \xi} = \frac{\partial F}{\partial x}\frac{\partial x}{\partial \xi} + \frac{\partial F}{\partial z}\frac{\partial z}{\partial \xi}$$

$$\frac{\partial F}{\partial \zeta} = \frac{\partial F}{\partial x}\frac{\partial x}{\partial \zeta} + \frac{\partial F}{\partial z}\frac{\partial z}{\partial \zeta} \tag{499}$$

which, when written in matrix notation, gives

$$\begin{bmatrix} \frac{\partial F}{\partial \xi} \\ \frac{\partial F}{\partial \zeta} \end{bmatrix} = \begin{bmatrix} \frac{\partial x}{\partial \xi} & \frac{\partial z}{\partial \xi} \\ \frac{\partial x}{\partial \zeta} & \frac{\partial z}{\partial \zeta} \end{bmatrix} \begin{bmatrix} \frac{\partial F}{\partial x} \\ \frac{\partial F}{\partial z} \end{bmatrix} \tag{500}$$

Hence,

$$\begin{bmatrix} \frac{\partial F}{\partial x} \\ \frac{\partial F}{\partial z} \end{bmatrix} = J^{-1} \begin{bmatrix} \frac{\partial F}{\partial \xi} \\ \frac{\partial F}{\partial \zeta} \end{bmatrix} \quad ; \quad J = \begin{bmatrix} \frac{\partial x}{\partial \xi} & \frac{\partial z}{\partial \xi} \\ \frac{\partial x}{\partial \zeta} & \frac{\partial z}{\partial \zeta} \end{bmatrix} \tag{501}$$

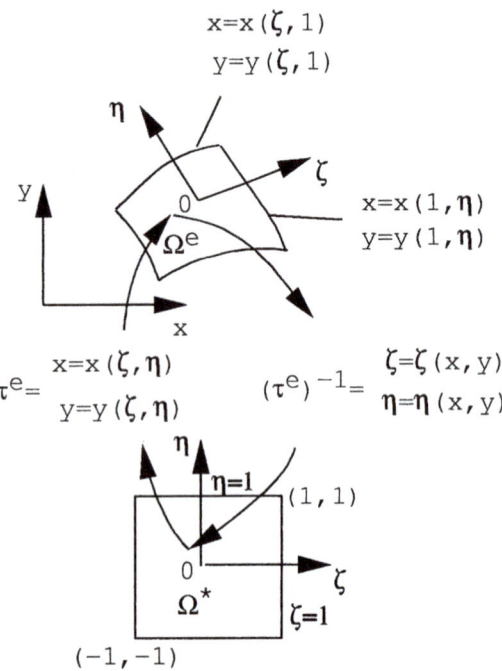

Fig. 146. Isoparametric mapping, notations

$$|J| = \begin{vmatrix} \frac{\partial x}{\partial \xi} & \frac{\partial z}{\partial \xi} \\[6pt] \frac{\partial x}{\partial \zeta} & \frac{\partial z}{\partial \zeta} \end{vmatrix} \tag{502}$$

where J is the Jacobian of the transformation and the following relation holds:

$$\iint\limits_{\Omega_e} dx\ dy = \int\limits_{-1}^{1}\int\limits_{-1}^{1} |J|\ d\xi\ d\zeta. \tag{503}$$

It follows from Eq. 500 and Eq. 503 that

$$\iint\limits_{\Omega^e} \frac{\partial F(\xi,\zeta)}{\partial x}\ d\Omega_u^e = \int\limits_{-1}^{1}\int\limits_{-1}^{1} \left(\bar{J}_{11} \frac{\partial F}{\partial \xi} + \bar{J}_{12} \frac{\partial F}{\partial \zeta} \right) |J|\ d\xi\ d\zeta \tag{504}$$

with $\bar{J}_{11}$ and $\bar{J}_{12}$ being the elements of the first row of the inverse matrix J^{-1}. On inverting the matrix J, the expression

$$J^{-1} = \frac{1}{|J|} \begin{bmatrix} \frac{\partial z}{\partial \zeta} & -\frac{\partial z}{\partial \xi} \\ -\frac{\partial x}{\partial \zeta} & \frac{\partial x}{\partial \xi} \end{bmatrix} \tag{505}$$

is obtained, and thus

$$\iint_{\Omega_u^e} \frac{\partial F(\xi,\zeta)}{\partial x} \, d\Omega_u^e = \int_{-1}^{1}\int_{-1}^{1} \frac{1}{|J|} \left(\frac{\partial F}{\partial \xi} \frac{\partial z}{\partial \zeta} - \frac{\partial F}{\partial \zeta} \frac{\partial z}{\partial \xi} \right) |J| \, d\xi \, d\zeta$$

$$= \int_{-1}^{1}\int_{-1}^{1} \left(\frac{\partial F}{\partial \xi} \frac{\partial z}{\partial \zeta} - \frac{\partial F}{\partial \zeta} \frac{\partial z}{\partial \xi} \right) \, d\xi \, d\zeta . \tag{506}$$

Owing to the properties of the basis functions $\{\Psi_{k,l}^{m,n}(\xi,\zeta)\}$ the transformations (τ_x^e, τ_z^e) have the following features:

- The mapping $\tau^e : \Omega^* \to \Omega^e$ is isomorphous, that is, every point $(x, z) \in \Omega^e$ is mapped into one and only one point $(\xi, \zeta) \in \Omega^*$ and the inverse relation holds $\tau^{e-1} : \Omega^e \to \Omega^*$, which is also unique.

- After the transformation, the nodes on the standard element correspond to the nodes on the element Ω^e.

Every piece of the boundary $\partial \Omega^*$ of Ω^* is mapped into the corresponding piece of the boundary $\partial \Omega^e$ of Ω^e defined by the equivalent nodal points and parameters prescribed on them. Applying the transformations described above, it is possible to map the approximating function $u^*(\xi, \zeta)$ and its domain of determinacy Ω^* into the function $u(x, z)$ and the domain Ω^e so that all the requirements concerning the C^1 continuous surface patch are satisfied, and the inverse transformations are possible.

5. Theoretical Wave Resistance Generated by a Ship Form

5.1. Quadrature for Outer Integral

To this end, $\cos \Theta = t$ is substituted into Eqs. 487 and 488, which results in the following expression

$$\Phi(u) = \frac{4\rho g^2}{\pi v_0^2} \int_0^1 (H_s^2 + H_c^2) \frac{dt}{t^3 \sqrt{1 - t^2}} \tag{507}$$

where

$$\left. \begin{array}{c} H_s \\ H_c \end{array} \right\} = \iint_{\Omega_u} \frac{\partial u(x,z)}{\partial x} \exp\left(\frac{-gz}{v_0^2 t^2} \right) \left\{ \begin{array}{c} \sin \\ \cos \end{array} \left(\frac{gx}{v_0^2 t} \right) \right. \, d\Omega_u . \tag{508}$$

The functional under consideration is a special case of the expression

$$J(x) = \int_0^1 (1 - x)^\alpha \, (1 + x)^\beta \, f(x) \, dx \quad ; \quad \text{where} \quad \alpha = \beta = -\frac{1}{2} \qquad (509)$$

The function $(1 - x^2)^{-\frac{1}{2}}$ is a weighting function, orthogonal to the Tchebyshev polynomials of the first kind: $T_N(x) = \cos(N \arccos x)$. Therefore, for this type of integral the optimum quadrature of the type

$$J(x) = \sum_{k=1}^N A_k \, f(x_k) \qquad (510)$$

is the following Gauss-Tchebyshev quadrature [38]

$$\int_0^1 \frac{f(x)dx}{\sqrt{1 - x^2}} = \frac{\pi}{2N} \sum_{k=1}^N f\left[\cos\left(\frac{2k - 1}{4N}\pi\right)\right] + R(f) \qquad (511)$$

where the approximation error is estimated by the formula

$$R(f) = \frac{\pi}{2^{2N-1}} \frac{f_{(\xi)}^{(2N)}}{2N!}, \quad 0 < \xi < 1. \qquad (512)$$

It follows from the above estimation that this approximation is accurate for polynomials of degree $\leq 2N - 1$. Study concerning the estimation of the numerical error in the above approximation [45], reveals satisfactory convergence for $N \sim 2^6$ dependent on the Froude number.

Making use of the above approximation, Eq. 507 can be written as follows

$$\Phi(u) \simeq \Phi^*(x) = 2\frac{\rho g^2}{v_0^2 N} \sum_{k=1}^N t_k^{-3} \, (H_s^2 + H_c^2) \qquad (513)$$

where

$$\left.\begin{array}{c} H_s \\ H_c \end{array}\right\} = \iint\limits_{\Omega^u} \frac{\partial u}{\partial x} \, \exp\left(\frac{-gz}{v_0^2 t_k^2}\right) \left\{\begin{array}{c} \sin \\ \cos \end{array}\right. \left(\frac{gx}{v_0^2 t_k}\right) d\Omega_u \qquad (514)$$

the nodes of the quadrature being determined from the equation

$$t_k = \cos\left(\frac{2k - 1}{4N}\pi\right) \quad ; \quad k = 1, 2, \ldots, N. \qquad (515)$$

The proposed approximation allows an analytic linear operator to be replaced with an algebraic operator which is convenient in further numerical computations, and where the Michell integral takes the form of a sum of N quadratic functionals.

5.2. Approximation of Inner Integrals

The inner integrals H_s and H_c express the magnitudes of the amplitudes of generated waves, dependent on the angle of their propagation $\Theta \in [0, \frac{\pi}{2}]$

$$
\left.\begin{array}{c} H_s \\ H_c \end{array}\right\} = \iint\limits_{\Omega} \frac{\partial u(x,z)}{\partial x} \exp\left(\frac{-gz}{v_0^2 t_k^2}\right) \left\{ \begin{array}{c} \sin \\ \cos \end{array} \right\} \left(\frac{gx}{v_0^2 t_k}\right) d\Omega_u. \tag{516}
$$

According to the assumed method of discretization of the domain Ω, the above integral is replaced with the sum of integrals over the individual domains Ω^e of the patches

$$
\left.\begin{array}{c} H_s \\ H_c \end{array}\right\} = \sum_{e=1}^{E} \iint\limits_{\Omega^e} \frac{\partial u^e(x,z)}{\partial x} \exp\left(\frac{-gz^e}{v_0^2 t_k^2}\right) \left\{ \begin{array}{c} \sin \\ \cos \end{array} \right\} \left(\frac{gx^e}{v_0^2 t_k}\right) d\Omega^e \tag{517}
$$

To simplify the notation, the following symbols are introduced

$$
\omega_k = \frac{-g}{v_0^2 t_k^2} \quad ; \quad \mu_k = \frac{g}{v_0^2 t_k} \tag{518}
$$

According to the assumed method of approximation on the elementary domain, formal substitutions are performed using the transformations given by Eqs. 497 and 498, with the following notation

$$
x^e(\xi,\zeta) = \Psi(\xi,\zeta; \{\hat{x}\}^e) = \sum_{l=1}^{n} N_l(\xi,\zeta)\, \hat{x}_l^e = \Psi^e(\xi,\zeta)
$$

$$
z^e(\xi,\zeta) = \phi(\xi,\zeta; \{\hat{z}\}^e) = \sum_{l=1}^{n} N_l(\xi,\zeta)\, \hat{z}_l^e = \phi^e(\xi,\zeta)
$$

$$
u^e(x,z) = u^e(x^e,z^e) = \Phi(\xi,\zeta; \{\hat{u}\}^e) = \sum_{l=1}^{n} N_l(\xi,\zeta)\, \hat{u}_l^e = \Phi^e(\xi,\zeta) \tag{519}
$$

for $e = 1,2,\ldots,E$. Points $(\xi,\zeta) \in \Omega^* \equiv [-1,1] \times [-1,1]$.

The relationship given by Eq. 506 is applied allowing the following transformation

$$
\begin{aligned}
\left.\begin{array}{c} H_s \\ H_c \end{array}\right\} &= \sum_{e=1}^{E} \int_{-1}^{1}\int_{-1}^{1} \frac{\partial\, \Phi^e(\xi,\zeta)}{\partial x} \exp[\omega_k\, \phi^e(\xi,\zeta)] \frac{1}{|J|} \left\{ \begin{array}{c} \sin \\ \cos \end{array} \right\} [\mu_k \Psi^e(\xi,\zeta)] \, |J| d\xi\, d\zeta \\[2mm]
&= \sum_{e=1}^{E} \int_{-1}^{1}\int_{-1}^{1} \left[\frac{\partial \Phi^e(\xi,\zeta)}{\partial \xi} \frac{\partial \phi^e(\xi,\zeta)}{\partial \zeta} - \frac{\partial \Phi^e(\xi,\zeta)}{\partial \zeta} \frac{\partial \phi^e(\xi,\zeta)}{\partial \xi} \right] \\[2mm]
&\qquad \cdot \exp[\omega_k\, \phi^e(\xi,\zeta)] \left\{ \begin{array}{c} \sin \\ \cos \end{array} \right\} [\mu_k\, \Psi^e(\xi,\zeta)] d\xi\, d\zeta \\[2mm]
&= \sum_{e=1}^{E} \int_{-1}^{1}\int_{-1}^{1} \left\{ \frac{\partial[\sum_{l=1}^{n} N_l(\xi,\zeta)\, \hat{u}_l^e]}{\partial \xi} \frac{\partial[\sum_{j=1}^{n} N_j(\xi,\zeta)\, \hat{z}_j^e]}{\partial \zeta} \right.
\end{aligned}
$$

$$- \frac{\partial[\sum_{l=1}^{n} N_l(\xi,\zeta) \, \hat{u}_l^e]}{\partial \zeta} \, \frac{\partial[\sum_{j=1}^{n} N_j(\xi,\zeta) \, \hat{z}_j^e]}{\partial \xi} \Bigg\}$$

$$\cdot \; \exp[\omega_k \sum_{l=1}^{n} N_l(\xi,\zeta) \, \hat{z}_l^e] \begin{Bmatrix} \sin \\ \cos \end{Bmatrix} [\mu_k \sum_{l=1}^{n} N_l(\xi,\zeta) \, \hat{x}_l^e] d\xi \; d\zeta. \tag{520}$$

The integrand in braces can be transformed to the equivalent form

$$\frac{\partial \Phi^e(\xi,\zeta)}{\partial x} = \sum_{l=1}^{n} \sum_{j=1}^{n} \left[\frac{\partial N_l(\xi,\zeta)}{\partial \xi} \, \frac{\partial N_j(\xi,\zeta)}{\partial \zeta} \, \hat{u}_l^e \, \hat{z}_j^e - \frac{\partial N_l(\xi,\zeta)}{\partial \zeta} \, \frac{\partial N_j(\xi,\zeta)}{\partial \xi} \, \hat{u}_l^e \, \hat{z}_j^e \right]$$

$$= \sum_{l=1}^{n} T_l(\xi,\zeta) \, \hat{u}_l^e \tag{521}$$

where

$$T_l^e(\xi,\zeta) = \sum_{j=1}^{n} \tau_{j,l}(\xi,\zeta) \, \hat{z}_j^e = \sum_{j=1}^{n} \left[\frac{\partial N_l(\xi,\zeta)}{\partial \xi} \, \frac{\partial N_j(\xi,\zeta)}{\partial \zeta} - \frac{\partial N_l(\xi,\zeta)}{\partial \zeta} \, \frac{\partial N_j(\xi,\zeta)}{\partial \xi} \right] \hat{z}_j^e \tag{522}$$

With such substitutions the integral considered here can be determined employing the equation

$$\begin{Bmatrix} H_s \\ H_c \end{Bmatrix} = \sum_{e=1}^{E} \sum_{l=1}^{n} \hat{u}_l^e \int_{-1}^{1} \int_{-1}^{1} T_l^e(\xi,\zeta) \, \exp\left[\omega_k \sum_{j=1}^{n} N_j(\xi,\zeta) \, \hat{z}_j^e\right]$$

$$\cdot \; \begin{Bmatrix} \sin \\ \cos \end{Bmatrix} \left[\mu_k \sum_{j=1}^{n} N_j(\xi,\zeta) \hat{x}_j^e \right] d\xi \; d\zeta. \tag{523}$$

The above integral is a nonelementary one and difficult to integrate numerically because of the multioscillating nature of the integrand which can change its sign as many as several hundred times (or more) on the domain of integration [45]. Therefore, the application of typical Gauss quadrature to determine this integral does not provide satisfactory convergence in a reasonable number of calculations.

The domain of integration on the standard element Ω^* is divided into subelements

$$\Omega_{s,t} \equiv [\xi_t, \xi_{t+1}] \times [\zeta_s, \zeta_{s+1}]. \tag{524}$$

The obtained partition is such that the standard element is the sum of elements

$$\Omega^* = \bigcup_{t=1}^{m-1} \bigcup_{s=1}^{m-1} \Omega_{s,t} = \bigcup_{t=1}^{m-1} \bigcup_{t=1}^{s=1} [\xi_t, \xi_{t+1}] \times [\zeta_s, \zeta_{s+1}] \tag{525}$$

where

$$\xi_t = \frac{2(t-1)}{m-2} - 1, \qquad t = 1, \cdots, m-1$$

$$\zeta_s = \frac{2(s-1)}{m-2} - 1, \qquad s = 1, \cdots, m-1. \tag{526}$$

On the standard element the arguments of the trigonometric functions and the exponential function as well as the function T_l^e are bicubic polynomials determining a certain surface on the domain Ω^*. It is assumed that on each subelement $\Omega_{s,t}$, the following approximation is performed with the notation

$$\left.\begin{array}{l} H_s \\ H_c \end{array}\right\} = \sum_{e=1}^{E} \sum_{l=1}^{n} \hat{u}_l^e \left\{\begin{array}{l} sh_l^e \\ ch_l^e \end{array}\right. \tag{527}$$

where

$$\left.\begin{array}{l} sh_l^e \\ ch_l^e \end{array}\right\} = \iint_{\Omega^*} T_l^e(\xi,\zeta)\exp\left[\omega_k \sum_{j=1}^{n} N_j(\xi,\zeta)\hat{z}_j^e\right]\left\{\begin{array}{l} \sin \\ \cos \end{array}\left[\mu_k \sum_{j=1}^{n} N_j(\xi,\zeta)x_l^e\right]d\xi\ d\zeta. \tag{528}$$

These integrals can be represented as the sum of integrals over the domains of the subelements

$$\left.\begin{array}{l} sh_l^e \\ ch_l^e \end{array}\right\} = \sum_{t=1}^{m-1} \sum_{s=1}^{m-1} \int_{\xi_t}^{\xi_{t+1}} \int_{\zeta_s}^{\zeta_{s+1}} \tau_{t,s}^{e,l}(\xi,\zeta)\exp\left[\omega_k\ \delta_{s,t}^e(\xi,\zeta)\right]\left\{\begin{array}{l} \sin \\ \cos \end{array}\left[\mu_k\kappa_{t,s}^e(\xi,\zeta)\right]d\xi\ d\zeta \tag{529}$$

On the subelement $\Omega_{s,t}$, the function $\tau_{t,s}^{e,l}$ is approximated with a bilinear function, and the functions $\delta_{t,s}^e$ and $\kappa_{t,s}^e$ are approximated with linear functions. Such an approach permits the separation of a surface integral into a product of two iterated integrals, each of the integrals being dependent only on one variable ξ or ζ. Consequently, the approximation of the integrand is obtained such that the integrals to be calculated can be approximated in closed form, which is exact for the approximating functions, and which produces the approximated function differing slightly from the exact function. The additional advantage is that the form of the equation is identical for both ξ and ζ variables and thus, only one procedure will suffice to compute the two integrals (integration with respect to ξ and ζ) which simplifies the development of a computer program.

For the bilinear form the following form of the approximating function is obtained

$$\begin{aligned} \tau_{t,s}^{e,l}(\xi,\zeta) &= \alpha_{t,s}^{e,l}\ \xi + \beta_{t,s}^{e,l}\ \zeta + \gamma_{t,s}^{e,l}\ \xi\zeta + \delta_{t,s}^{e,l} = \\ &= (A_{t,s}\xi + B_{t,s})(A_{t,s}^{e,l}\ \zeta + B_{t,s}^{e,l}) + (a_{t,s}\ \xi + b_{t,s})(a_{t,s}^{e,l}\ \zeta + b_{t,s}^{e,l}) \\ &= \left[\frac{\xi - \xi_t}{(\xi_{t+1} - \xi_t)(\zeta_{s+1} - \zeta_s)}\right]\left[(\eta_{t+1,s+1} - \eta_{t+1,s})\zeta \right. \\ &\left. +\ \zeta_{s+1}\eta_{t+1,s} - \zeta_s\eta_{t+1,s+1}\right] + \left[\frac{\xi - \xi_{t+1}}{(\xi_{t+1} - \xi_t)(\zeta_{s+1} - \zeta_s)}\right] \\ &\cdot \left[(\eta_{t,s} - \eta_{t,s+1})\zeta + \zeta_s\eta_{t,s+1} - \zeta_{s+1}\eta_{t,s}\right]. \end{aligned} \tag{530}$$

At the four nodes of the subelement the interpolation condition holds

$$\tau_{t,s}^{e,l}(\xi, \zeta) = T_l^e(\xi_t, \zeta_s).$$ (531)

The linear approximations are obtained by taking the linear terms of the Taylor series expansion

$$\delta_{t,s}^e(\xi, \zeta) = \delta^e(\xi_t, \zeta_s) + \frac{\partial \delta^e(\xi, \zeta)}{\partial \xi}\Big|_{\xi_t, \zeta_s}(\xi - \xi_t) + \frac{\partial \delta^e(\xi, \zeta)}{\partial \zeta}\Big|_{\xi_t, \zeta_s}(\zeta - \zeta_t).$$ (532)

Upon replacing the derivatives with the differential quotient, the expression is obtained

$$\begin{aligned}
\delta_{t,s}^e(\xi, \zeta) &= \delta^e(\xi_t, \zeta_s) + \frac{\delta^e(\xi_{t+1}, \zeta_s) - \delta^e(\xi_t, \zeta_s)}{\xi_{t+1} - \xi_t}(\xi - \xi_t) \\
&+ \frac{\delta^e(\xi_t, \zeta_{s+1}) - \delta^e(\xi_t, \zeta_s)}{\zeta_{s+1} - \zeta_s}(\zeta - \zeta_s)
\end{aligned}$$ (533)

where

$$\delta^e(\xi, \zeta) = \sum_{j=1}^n N_j(\xi, \zeta)\, \hat{z}_j^e.$$ (534)

The following notation is introduced

$$\begin{aligned}
E_{t,s}^e &= \frac{\delta^e(\xi_{t+1}, \zeta_s) - \delta^e(\xi_t, \zeta_s)}{\xi_{t+1} - \xi_t} \\
G_{t,s}^e &= \frac{\delta^e(\xi_t, \zeta_{s+1}) - \delta^e(\xi_t, \zeta_s)}{\zeta_{s+1} - \zeta_s} \\
C_{t,s}^e &= \delta^e(\xi_t, \zeta_s) - E_{t,s}^e\, \xi_t - G_{t,s}^e \zeta_s.
\end{aligned}$$ (535)

With the above notation the final form of the formula is as follows

$$\delta_{t,s}^e(\xi, \zeta) = \left(E_{t,s}^e \xi + \frac{1}{2}C_{t,s}^e\right) + \left(G_{t,s}^e \zeta + \frac{1}{2}C_{t,s}^e\right)$$ (536)

where $(\xi, \zeta) \in [\xi_t, \xi_{t+1}] \times [\zeta_s, \zeta_{s+1}]$. By analogy, the following expression are obtained for the function $\kappa_{t,s}^e(\xi, \zeta)$.

The following notation is introduced

$$\begin{aligned}
W_{t,s}^e &= \frac{\kappa^e(\xi_{t+1}, \zeta_s) - \kappa^e(\xi_t, \zeta_s)}{\xi_{t+1} - \xi_t} \\
V_{t,s}^e &= \frac{\kappa^e(\xi_t, \zeta_{s+1}) - \kappa^e(\xi_t, \zeta_s)}{\zeta_{s+1} - \zeta_s} \\
S_{t,s}^e &= \kappa^e(\xi_t, \zeta_s) - W_{t,s}^e\xi_t - V_{t,s}^e\zeta_s
\end{aligned}$$ (537)

where

$$\kappa^e(\xi_t, \zeta_s) = \sum_{j=1}^{n} N_j(\xi_t, \zeta_s)\hat{x}_j^e.$$

The final formula takes the form

$$\kappa_{t,s}^e(\xi, \zeta) = (W_{t,s}^e\xi + \frac{1}{2}S_{t,s}^e) + (V_{t,s}^e\zeta + \frac{1}{2}S_{t,s}^e)|_{(\xi,\zeta)\in\Omega_{t,s}}. \tag{538}$$

Returning to the integrals expressed by Eq. 529 and using the above approximations the following relations are obtained

$$\left.\begin{matrix}sh_l^e\\ch_l^e\end{matrix}\right\} = \sum_{t=1}^{m-1}\sum_{s=1}^{m-1}\left\{\begin{matrix}sh_{t,s}^{e,l}\\ch_{t,s}^{e,l}\end{matrix}\right. \tag{539}$$

where

$$\left.\begin{matrix}sh_{t,s}^{e,l}\\ch_{t,s}^{e,l}\end{matrix}\right\} = \int_{\xi_t}^{\xi_{t+1}}\int_{\zeta_s}^{\zeta_{s+1}}\Big\{\Big[(A_{t,s}\,\xi + B_{t,s})(A_{t,s}^{e,l}\zeta + B_{t,s}^{e,l})$$

$$+ \ (a_{t,s}\xi + b_{t,s})(a_{t,s}^{e,l}\zeta + b_{t,s}^{e,l})\Big]$$

$$\cdot \ \exp\Big[\omega_k(E_{t,s}^e\xi + \frac{1}{2}C_{t,s}^e) + \omega_k(G_{t,s}^e\zeta + \frac{1}{2}C_{t,s}^e)\Big]$$

$$\cdot \ \left\{\begin{matrix}\sin\\\cos\end{matrix}\right. [\mu_k(W_{t,s}^e\xi + \frac{1}{2}S_{t,s}^e) + \mu_k(V_{t,s}^e\zeta + \frac{1}{2}S_{t,s}^e)]\Big\}d\xi d\zeta$$

$$= \int_{\xi_t}^{\xi_{t+1}}\int_{\zeta_s}^{\zeta_{s+1}}\Big[F_{x_{t,s}}^{e,l}(\xi)F_{z_{t,s}}^{e,l}(\zeta) + f_{x_{t,s}}^{e,l}(\xi)f_{z_{t,s}}^{e,l}(\zeta)\Big]$$

$$\cdot \ \exp\Big[\phi_{x_{t,s}}^e(\xi) + \phi_{z_{t,s}}^e(\zeta)\Big]\left\{\begin{matrix}\sin\\\cos\end{matrix}\right. \Big[\Psi_{x_{t,s}}^e(\xi) + \Psi_{z_{t,s}}^e(\zeta)\Big]d\xi\ d\zeta. \tag{540}$$

Using the following relations

$$\begin{aligned}\exp[\phi_{x_{t,s}}^e(\xi)\phi_{z_{t,s}}^e(\zeta)] &= \exp\phi_{x_{t,s}}^e(\xi) + \exp\phi_{z_{t,s}}^e(\zeta)\\\sin[\Psi_{x_{t,s}}^e(\xi) + \Psi_{z_{t,s}}^e(\zeta)] &= \sin\Psi_{x_{t,s}}^e(\xi)\cos\Psi_{z_{t,s}}^e(\zeta)\\&+ \cos\Psi_{x_{t,s}}^e(\xi)\sin\Psi_{z_{t,s}}^e(\zeta)\\\cos[\Psi_{x_{t,s}}^e(\xi) + \Psi_{z_{t,s}}^e(\zeta)] &= \cos\Psi_{x_{t,s}}^e(\xi)\cos\Psi_{z_{t,s}}^e(\zeta)\\&- \sin\Psi_{x_{t,s}}^e(\xi)\sin\Psi_{z_{t,s}}^e(\zeta)\end{aligned} \tag{541}$$

Eq. 540 can be expressed in the following form

$$\left.\begin{matrix}sh_{t,s}^{e,l}\\ch_{t,s}^{e,l}\end{matrix}\right\} = \int_{\xi_t}^{\xi_{t+1}}\int_{\zeta_s}^{\zeta_{s+1}}F_{x_{t,s}}^{e,l}F_{z_{t,s}}^{e,l}\exp(\phi_{x_{t,s}}^e)\exp(\phi_{z_{t,s}}^e)\left\{\begin{matrix}\sin\\\cos\end{matrix}\right. (\Psi_{x_{t,s}}^e)$$

$$\cdot \left\{ \begin{matrix} \cos \\ \cos \end{matrix} \right. (\Psi^e_{z_{t,s}}) d\xi \; d\zeta + \int_{\xi_t}^{\xi_{t+1}} \int_{\zeta_s}^{\zeta_{s+1}} f^{e,l}_{x_{t,s}} f^{e,l}_{z_{t,s}} \exp(\phi^e_{x_{t,s}}) \exp(\phi^e_{z_{t,s}})$$

$$\cdot \left\{ \begin{matrix} \sin \\ \cos \end{matrix} \right. (\Psi^e_{x_{t,s}}) \left\{ \begin{matrix} \cos \\ \cos \end{matrix} \right. (\Psi^e_{z_{t,s}}) d\xi \; d\zeta + \int_{\xi_t}^{\xi_{t+1}} \int_{\zeta_s}^{\zeta_{s+1}} F^{e,l}_{x_{t,s}} F^{e,l}_{z_{t,s}}$$

$$\cdot \exp(\phi^e_{x_{t,s}}) \exp(\phi^e_{z_{t,s}}) \left\{ \begin{matrix} \cos \\ -\sin \end{matrix} \right. (\Psi^e_{x_{t,s}}) \left\{ \begin{matrix} \sin \\ \sin \end{matrix} \right. (\Psi^e_{z_{t,s}}) d\xi \; d\zeta$$

$$+ \int_{\xi_t}^{\xi_{t+1}} \int_{\zeta_s}^{\zeta_{s+1}} f^{e,l}_{x_{t,s}} f^{e,l}_{z_{t,s}} \exp(\phi^e_{x_{t,s}}) \exp(\phi^e_{z_{t,s}})$$

$$\cdot \left\{ \begin{matrix} \cos \\ -\sin \end{matrix} \right. (\Psi^e_{x_{t,s}}) \left\{ \begin{matrix} \sin \\ \sin \end{matrix} \right. (\Psi^e_{z_{t,s}}) d\xi \; d\zeta. \tag{542}$$

Replacing the double integrals with products of single integrals the final form of Eq. 540 is obtained

$$\left. \begin{matrix} sh^{e,l}_{t,s} \\ ch^{e,l}_{t,s} \end{matrix} \right\} = \int_{\xi_t}^{\xi_{t+1}} F^{e,l}_{x_{t,s}} \exp(\phi^e_{x_{t,s}}) \left\{ \begin{matrix} \sin \\ \cos \end{matrix} \right. (\Psi^e_{x_{t,s}}) d\xi$$

$$\cdot \int_{\zeta_s}^{\zeta_{s+1}} F^{e,l}_{z_{t,s}} \exp(\phi^e_{z_{t,s}}) \left\{ \begin{matrix} \cos \\ \cos \end{matrix} \right. (\Psi^e_{z_{t,s}}) d\zeta$$

$$+ \int_{\xi_t}^{\xi_{t+1}} f^{e,l}_{x_{t,s}} \exp(\phi^e_{x_{t,s}}) \left\{ \begin{matrix} \sin \\ \cos \end{matrix} \right. (\Psi^e_{x_{t,s}}) d\xi$$

$$\cdot \int_{\zeta_s}^{\zeta_{s+1}} f^{e,l}_{z_{t,s}} \exp(\phi^e_{z_{t,s}}) \left\{ \begin{matrix} \cos \\ \cos \end{matrix} \right. (\Psi^e_{z_{t,s}}) d\zeta$$

$$\left\{ \begin{matrix} + \\ - \end{matrix} \right. \int_{\xi_t}^{\xi_{t+1}} F^e_{x_{t,s}} \exp(\phi^e_{x_{t,s}}) \left\{ \begin{matrix} \cos \\ \sin \end{matrix} \right. (\Psi^e_{x_{t,s}}) d\xi$$

$$\cdot \int_{\zeta_s}^{\zeta_{s+1}} F^{e,l}_{z_{t,s}} \exp(\phi^e_{z_{t,s}}) \left\{ \begin{matrix} \sin \\ \sin \end{matrix} \right. (\Psi^e_{z_{t,s}}) d\zeta$$

$$\left\{ \begin{matrix} + \\ - \end{matrix} \right. \int_{\xi_t}^{\xi_{t+1}} f^{e,l}_{x_{t,s}} \exp(\phi^e_{x_{t,s}}) \left\{ \begin{matrix} \cos \\ \sin \end{matrix} \right. (\Psi^e_{x_{t,s}}) d\xi$$

$$\cdot \int_{\zeta_s}^{\zeta_{s+1}} f^{e,l}_{z_{t,s}} \exp(\phi^e_{z_{t,s}}) \left\{ \begin{matrix} \sin \\ \sin \end{matrix} \right. (\Psi^e_{z_{t,s}}) d\zeta. \tag{543}$$

Each of the above single integrals takes one of the forms

$$\left.\begin{matrix} J_s \\ J_c \end{matrix}\right\} = \int\limits_{y1}^{y2} (\alpha y + \beta) \exp(\gamma y + \delta) \left\{ \begin{matrix} \sin \\ \cos \end{matrix} \right. (\epsilon y + \rho) dy \tag{544}$$

which can be expressed by the exact form

$$\left.\begin{matrix} J_s \\ J_c \end{matrix}\right\} = \frac{\exp(\gamma\, y + \delta)}{\gamma^2 + \epsilon^2} \left\{ \left\{ \begin{matrix} \sin \\ \cos \end{matrix} \right. (\epsilon\, y + \rho) \left[\gamma(\alpha\, y + \beta) - \frac{\alpha(\gamma^2 - \epsilon^2)}{\gamma^2 + \epsilon^2} \right] \right\}$$

$$+ \left. \left\{ \left\{ \begin{matrix} -\cos \\ \sin \end{matrix} \right. (\epsilon\, y + \rho)[\epsilon(\alpha\, y + \beta) - \frac{2\alpha\,\gamma\,\epsilon}{\gamma^2 + \epsilon^2}] \right\} \right|_{y1}^{y2}. \tag{545}$$

Finally, if t denotes the time consumed by a computer to calculate this expression, then the time T needed to determine the value of the expression for a typical example, for which the following constants can be assumed

$$\begin{matrix} \min & K \sim 32 \\ \min & E \sim 9 \\ & n = 16 \\ \min & m \sim 5 \end{matrix}$$

is of the order

$$T = \sum_{k=1}^{K} \sum_{e=1}^{E} \sum_{i=1}^{m} \sum_{t=1}^{m-1} \sum_{s=1}^{m-1} (2 \cdot 16 \cdot t) = 32 \cdot K \cdot E \cdot n(m-1)^2 t = 2359396t \tag{546}$$

This estimation shows that it is justified to employ a computer of big computational power for solving the problem under consideration.

BIBLIOGRAPHY

1. I. H. Abbott and A. E. von Doenhoff, Theory of Wing Sections (McGraw-Hill Book Company, Inc. 1949).

2. B. Barsky, Rational Beta-Splines for Representing Curves and Surfaces, *IEEE Computer Graphics and Applications*, 1993.

3. R. Bartels, B. Barsky and J. Beatty, *An Introduction to Splines for Use in Computer Graphics and Geometric Modeling* (Morgan Kaufman Publishers, Inc. 1987).

4. P. Bézier, *Emploi des machines a commande numérique* (Masson and Cie, Paris, 1970).

5. P. Bézier, Courbes et surfaces, *Mathématiques et CAO* **4** (Hermès Publ., Paris, 1986).

6. M. I. G. Bloor and M. J. Wilson, Generating Blend Surfaces using Partial Differential Equations, *CAD* **21**, 1989.

7. M. I. G. Bloor and M. J. Wilson, Blend Design as a Boundary-Value Problem, *Geometric Modelling: Theory and Practice* (Springer-Verlag, Berlin, 1989).

8. M. I. G. Bloor and M. J. Wilson, Using Partial Differential Equations to Generate Free-Form Surfaces, *CAD* **22**, 1990.

9. W. Boehm, Inserting New Knots into B-Spline Curves, *CAD* **12**, 1980.

10. W. Boehm, G. Farin and J. Kahmann, A Survey of Curve and Surface Methods in CAGD, *CAGD* (North-Holland, 1984).

11. J. M. Brown, *The Design and Properties of Surfaces Generated using Partial Differential Equations*, Ph.D Thesis, Univ of Leeds, Leeds, U. K. 1992.

12. L. Buczkowski, Mathematical Construction, Approximation, and Design of the Ship Body Form, *Journal of Ship Research*, 1969.

13. M. P. do Carmo, *Differential Geometry of Curves and Surfaces* (Prentice-Hall, Inc. 1976).

14. F. H. Chapman, A Treatise on Ship-Building, 1760.

15. S. Y. Cheng, Blending and Fairing using Partial Differential Equations, Ph.D Thesis, Univ of Leeds, Leeds, U. K. 1992.

16. H. Chiyokura and F. Kimura, Design of Solids with Free-Form Surfaces, *ACM Computer Graphics* **17**, 1983.

17. L. Collatz, *The Numerical Treatment of Differential Equations* (Springer-Verlag, 1960).

18. S. A. Coons, Surfaces for Computer Aided Design of Space Forms, *Report: MAC-TR-41*, (M. I. T., Mechanical Eng. Dept, 1964).

19. E. Cohen, T. Lyche and R. Riesenfeld, Discrete B-Splines and Subdivision Techniques in Computer-Aided Geometric Design and Computer Graphics, *Comp. Graph. and Image Proc.* 14, 1980.

20. E. Cohen, T. Lyche and L. L. Schumaker, Algorithms for Degree Raising of Splines, *ACM Transaction on Graphics* 4, 1985.

21. C. W. Dekanski, *Design and Analysis of Propeller Blade Geometry using the PDE Method*, Ph.D Thesis, Univ of Leeds, Leeds, U. K. 1993.

22. J. C. Dill, An Application of Color Graphics to the Display of Surface Curvature, *Computer Graphics* 15, 1981.

23. Q. Ding and B. J. Davies, *Surface Engineering Geometry for Computer-Aided Design and Manufacture* (J. Wiley and Sons, 1987).

24. G. Farin, A Construction for Visual Continuity of Polynomial Surface Patches, *Computer Graphics and Image Processing*, 1982.

25. G. Farin, *Curves and Surfaces for Computer Aided Geometric Design, A Practical Guide* (Academic Press, San Diego, 1990).

26. I. D. Faux and M. J. Pratt, *Computational Geometry for Design and Manufacture* (Ellis Horwood, 1979).

27. W. J. Gordon and R. Riesenfeld, B-Spline Curves and Surfaces, *CAGD* (Academic Press, 1974).

28. J. A. Gregory, Smooth Interpolation with Twist Constraints, *Computer Aided Geometric Design* (Academic Press, New York, 1974).

29. H. Hagen and G. Schulze, Automatic Smoothing with Geometric Surface Patches, *Computer Aided Geometric Design* 4, 1987.

30. J. C. Holladay, Smoothest Curve Approximation, *Math Tables Aids Computations* 11, 1957.

31. M. Hosaka, Theory of Curve and Surface Synthesis and their Smooth Fitting, *Information Processing in Japan* 9, 1969.

32. M. Hosaka and F. Kimura, Synthesis Method of Curve and Surface in Interactive CAD, *Interactive Techniques in CAD* (IEEE, New York, 1978).

33. J. Hoschek and D. Lasser, *Fundamentals of Geometric Data Processing*, (in German), (B. G. Teuber Verlag, Stuttgart, 1992).

34. M. Jankowski, *Elementy Grafiki Komputerowej* (in Polish) (WNT, Warszawa, 1990).

35. P. D. Kaklis and H. Nowacki, Experiences in Curve and Surface Fairing, *Symposium on Computer Geometry, Gaussig, TU Dresden*, 1990.

36. M. Kallay and B. Ravani, Optimal Twist Vector as a Tool for Interpolating a Network of Curves with a Minimum Energy Surface, *Computer Aided Geometric Design* **7**, 1990.

37. A. A. Kostiukov, *Teoria Korabielnych Voln i Volnovovo Soprotivlienia* (Sudprangiz, Leningrad, 1959).

38. J. V. Krylov, *Approximate Calculation of Integrals* (The Macmillan Company, New York, 1962).

39. G. Kuiper, Preliminary Design of Ship Lines by Mathematical Methods, *Journal of Ship Research* **14**, 1970.

40. H. Lackenby, On the Systematic Geometrical Variation of Ship Forms, *Trans INA* **92**, London, 1950.

41. D. Liu and J. Hoschek, GC^1 Continuity Conditions between Adjacent Rectangular and Triangular Bézier Surface Patches, *CAD* **21**,1989.

42. T. Lyche and K. Morken, Knot Removal for Parametric B-Spline Curves and Surfaces, *Computer Aided Geometric Design* **4**, 1987.

43. H. Meier, *Differential Geometry Based Design and Analytic Representation of Curvature Continuous Ship Surfaces and Similar Free Form Surfaces* (VDI-Verlag, Reihe Rechnerunterstüzte Verfahren, Dusseldorf) **5** (in German).

44. H. Meier and H. Nowacki, Interpolating Curves with Gradual Changes in Curvature, *Computer Aided Geometric Design* **4**, 1987.

45. J. P. Michalski, *Investigations of the Fitness of Chosen Quadratures to the Calculation of Wave Resistance of Ship* (Zeszyty Naukowe Politechniki Gdańskiej, Budownictwo Okretowe, 1974).

46. J. H. Michell, The Wave Resistance of a Ship, *Philosophical Magazine* **45**, 1898.

47. M. E. Mortenson, *Geometric Modeling* (Wiley-Interscience, 1985).

48. F. Munchmeyer, Shape Interrogation: A Case Study. Geometric Modeling Algorithms and New Trends, *SIAM*, 1987.

49. H. Nowacki, Recent Development in Computer-Aided Sculptured Surface Modelling, *Symposium on Computer-Aided Hull Surface Definition, Annapolis, Md.*, 1977.

50. H. Nowacki, Curve and Surface Generation and Fairing, *Computer Aided Design, Modelling, Systems Engineering, CAD Systems, Lecture Notes in Computer Science* (Springer-Verlag, Berlin, 1980).

51. H. Nowacki, Mathematical Methods for Fairing of Curves and Surfaces, *Geometric Procedures of Graphical Data Processing* (Springer-Verlag, Berlin, Heidelberg, 1990) (in German).

52. H. Nowacki, G. Creutz and F. Munchmeyer, Ship Lines Creation by Computer - Objectives, Methods and Results, *Symposium on Computer-Aided Hull Surface Defintion, Society of Naval Architects and Marine Engineers, Annapolis*, 1977.

53. H. Nowacki, P. D. Kaklis and J. Weber, Curve Mesh Fairing and GC^2 Surface Interpolation, *Modélisation mathématique et analyse numérique* **26** (Afcet, Gauthier-Villars, 1992).

54. H. Nowacki, D. Liu and X. Lü, Mesh Fairing GC^1 Surface Generation Method, *Theory and Practice of Geometric Modelling* (Springer-Verlag, Berlin, Heidelberg, 1989).

55. H. Nowacki, D. Liu and X. Lü, Fairing Bézier Curves with Constraints, *Computer Aided Geometric Design* **7**, 1990.

56. H. Nowacki and D. Reese, Design and Fairing of Ship Surfaces, *Surfaces in CAGD* (North-Holland Publ., Amsterdam, 1983).

57. J. T. Oden and G. F. Carey, *Finite Elements Mathematical Aspects* (Prentice-Hall Inc. New Jersey, 1983).

58. J. M. Ortega and W. G. Poole, *An Introduction to Numerical Methods for Differential Equations* (Pitman Publishing Inc. 1981).

59. L. Piegl, On NURBS: A Survey, *IEEE Computer Graphics and Applications*, 1991.

60. L. Piegl and W. Tiller, Curve and Surface Constructions using Rational B-splines, *CAD*, 1987.

61. D. Pilcher, Smooth Parametric Surfaces, *Computer Aided Geometric Design* (Academic Press, New York, 1974).

62. H. Prautzsch, Degree Elevation of B-Spline Curves, *Computer Aided Geometric Design* **1**, 1984.

63. H. Prautzsch and B. Piper, A Fast Algorithm to Raise the Degree of Spline Curves, *Computer Aided Geometric Design* **8**, 1991.

64. P. M. Prenter, *Splines and Variational Methods* (J. Wiley and Sons, 1985).

65. H. W. Press, *Numerical Recipes. The Art of Scientific Computing* (Cambridge University Press, 1989).

66. K. Puchstein, Possibilities in Changing Ship Form Parameters by Geometric Distortion of the Hull, *Schiffbauforschung*, (in German), 1965.

67. U. Rabien, Computer-Aided Design of Ship Lines, Departing from Basic Forms, *STG Publication by Panel on Ship Design and Safety*, Hamburg, (in German), 1975.

68. H. E. Saunders, Hydrodynamics in Ship Design, *Trans. Society of Naval Architects and Marine Engineers* **2**, 1957.

69. D. R. Smith, *The Geometry of Marine Propellers* (Defence Research Establishment Atlantic, 1988).

70. G. D. Smith, *Numerical Solutions of Partial Differential Equations* (Oxford University Press, 1971).

71. N. B. Standerski, *The Generation and Distortion of Ship Surfaces Represented by Global Tensor Product Surfaces*, (in German), Dissertation, TU Berlin, 1988.

72. G. Strang and G. J. Fix, *An Analysis of the Finite Element Method* (Prentice-Hall Inc. New Jersey, 1973).

73. D. J. Struik, *Lectures on Classical Differential Geometry* (Dover Publications, Inc. 1961).

74. D. W. Taylor, Calculations of Ships' Forms and Light Thrown by Model Experiments upon Resistance, Propulsion and Rolling of Ships, *Intl. Congress of Engineering, San Francisco*, 1915.

75. W. Tiller, Rational B-Splines for Curve and Surface Representation, *IEEE Computer Graphics and Applications*, 1983.

76. U. Umlauf, *Approximately Curvature Continuous Transitions between Free-Form Surfaces*, Dissertation, TU Berlin, 1993.

77. K. J. Versprille, *Computer-Aided Design Applications of the Rational B-Spline Approximation Form*, Ph.D Thesis, Sracuse Univ, Syracuse N. Y. 1975.

78. H. Walter, *Numerical Representation of Surfaces Based on an Optimization Principle*, Dissertation, (in German) TU Munich, 1971.

79. J. Weber, *Methods for Constructing Curvature Continuous Free-Form Surfaces*, (in German), Dissertation, TU Berlin, 1990.

80. G. Weinblum, *Systematic Development of Hull Forms*, (in German), Trans. STG, 1953.

81. C. D. Woodward, Methods for Cross-Sectional Design of B-Spline Surfaces, *EU-ROGRAPHICS '86* (North Holland Publ., Amsterdam, 1986).

82. C. D. Woodward, Skinning Techniques for Interactive B-Spline Surface Interpolation, *CAD* **20**, 1988.

83. F. Yamaguchi, *Curves and Surfaces in Computer Aided Geometric Design* (Springer-Verlag, 1988).

84. X. Ye, *Construction and Verification of Smooth Free-Form Surfaces Generated by Compatible Interpolation of Arbitrary Meshes*, Dissertation, TU Berlin, 1994.

85. O. C. Zienkiewicz, *Metoda Elementów Skończonych* (Arkady, Warszawa, 1972).

INDEX

CPSIA information can be obtained
at www.ICGtesting.com
Printed in the USA
JSHW050939230721
17159JS00001B/78